Materials and the designer

Materials

and the designer

E. H. CORNISH

Formerly Manager Materials Projects, Standard Telecommunication Laboratories Ltd

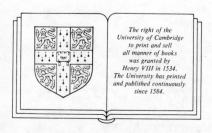

The right of the University of Cambridge to print and sell all manner of books was granted by Henry VIII in 1534. The University has printed and published continuously since 1584.

CAMBRIDGE UNIVERSITY PRESS

Cambridge

London   New York   New Rochelle

Melbourne   Sydney

Published by the Press Syndicate of the University of Cambridge
The Pitt Building, Trumpington Street, Cambridge CB2 1RP
32 East 57th Street, New York, NY 10022, USA
10 Stamford Road, Oakleigh, Melbourne 3166, Australia

First published 1987

Printed in Great Britain at the University Press, Cambridge

British Library cataloguing in publication data
Cornish, Eric H.
  Materials and the designer.

  1. Materials   2. Design, Industrial
  I. Title
  620.1′1   TA403.6

Library of Congress cataloguing in publication data
Cornish, E. H.
  Materials and the designer.

  Bibliography.
  Includes index.
  1. Materials.   2. Engineering design.   I. Title.
TA403.C644   1986   620.1′1   86-6087

ISBN 0 521 30734 1

Contents

Preface   x

Introduction   1
The need to know   1
The product in its environment   2
The materials cycle   3
The role of processing   4
Materials' structures   5
Selecting materials   6

1   The impact of design on manufacturing industry   7
1.1   The role of design in product manufacturing   7
1.2   The designer's contribution to the requirements of industry   11
1.3   Management functions associated with design   14
1.4   Design activity in a Company   15
1.5   The designer as a focus of knowledge   17
1.6   Attributes of good product design   19
1.7   Materials as a factor in design   21

2   Expertise required for the design process   23
2.1   The importance of mechanical design   23
2.2   Cost-effective materials and manufacturing processes   24
2.3   The challenge of materials selection   25
2.4   Materials aspects of design   27
2.5   Data bases and computer aids for materials selection   31

3   An introduction to materials   34
3.1   Properties and performance limitations   38
3.2   Salient features of main classes of materials   43

4   Properties of metals and alloys   55
4.1   Aspects of selecting metals   59
4.2   Characteristics of specific metals   67
4.3   Welding   73

4.4	Machining	74
4.5	Surface conversion	78
5	**Properties of ceramics**	79
5.1	General characteristics	79
5.2	Varieties available	80
5.3	Comparative properties	82
5.4	Data on some specific ceramics	85
5.5	New developments in ceramics and glass	87
5.6	Fabrication processes	89
5.7	Design factors	90
6	**Properties of polymers**	91
6.1	Types and varieties available	93
6.2	Elastomers	95
6.3	Cellular materials	96
6.4	Plastic film	97
6.5	Costs relative to other materials	97
6.6	General properties of plastic materials	99
6.7	Some specific aspects of plastics	105
6.8	Fabrication routes	113
6.9	Some aspects of plastics selection and design	119
6.10	Some recent developments in polymers	122
7	**Properties of composites**	124
7.1	Definition of composite types	124
7.2	Reinforcements in common use	127
7.3	Advantages and limitations of composites	129
7.4	Design considerations	134
7.5	Production techniques	137
7.6	Reinforced plastics	139
8	**Materials' performance in service**	143
8.1	Metals	144
8.2	Ceramics	148
8.3	Polymers	151
8.4	Composites	161
8.5	Flammability	164
9	**Finishes and coatings as protective systems**	182
9.1	Characteristics of surface coatings	187
9.2	Selection of finishes	188
9.3	Finishes for modifying surfaces	193
9.4	Metal based surface layers	196
9.5	Ceramic finishes	204
9.6	Organic (paint) finishes	205

10   Materials reliability and service life   208
10.1   Predicting service life   208
10.2   Design decisions   212
10.3   The lessons of failure in service   213
10.4   Failure of materials   214
10.5   Product quality   216
10.6   Use of statistics   218

11   Factors controlling the selection of substitute materials   219
11.1   Materials are not totally interchangeable   219
11.2   Reasons for substitution   220
11.3   Product improvement   223
11.4   Success and failure   224
11.5   Strategic materials and economics   225
11.6   Effects on production processes   228

12   Material forming processes and design   230
12.1   Metals   230
12.2   Polymers   235
12.3   Composites   237
12.4   Ceramics   238
12.5   Newer production possibilities   239

13   Sources of information on materials   241
13.1   Introduction   241
13.2   Location of documents   241
13.3   Published information   242
13.4   Current research   252
13.5   Further information sources   256
13.6   Materials suppliers   259

14   Standards and materials   261
14.1   Types of standards   261
14.2   Standards available   262
14.3   Codes of Practice   263
14.4   References on Standards   263
       References   265
       Bibliography   272
       Index   274

Preface

This book does not aspire to discuss pure materials science topics, nor to offer comprehensive numerical design data for each material, although reference is made to underlying materials science concepts wherever opportune. It does, however, seek to identify those parameters which must be considered when selecting materials for use in engineering applications. Emphasis is put on the needs of manufacturing industry, which is defined for present purposes as involved in production of prefabricated parts and components intended for civil, domestic, marine, military, aerospace and chemical processing applications. The decision has been taken to omit all aspects of the building and construction industry, so no reference is made to concrete, timber, bitumens, soil and masonry.

The chapter on the impact of design on manufacturing industry is intended to demonstrate how correct selection of materials in terms of their performance, availability and cost, together with exploitation of available production capability, will enhance the profitability of a commercial operation.

Having embarked upon a design study, it is necessary for the expertise and judgement of the design engineer to be coordinated with those of other experts to produce the most effective result. The chapter on the design process draws attention to the variety of knowledge and advice which the designer needs to draw upon to make materials choices with a high level of confidence.

A large number of metallic and non-metallic materials, each with a wide spectrum of properties, is available to the designer. The salient features of the main materials families, and their typical properties and performance limitations in service are outlined as a guide to further, in-depth, study. Surface coatings and finishes, as systems for protection in adverse service environments, are also considered.

Guidelines by which degradation and failure mechanisms, induced by chemical and mechanical agencies, can be identified and assessed to provide economic safety margins in use, are briefly presented, and existing world standards for quality assurance and product control are reviewed.

Factors controlling the selection of alternative materials in an existing product design receive attention, being perhaps necessitated by new market forces as a result of changes in service requirements or in production processes. Some economic and strategic factors related to the supply of raw materials are discussed, and a concept of an 'envelope of properties' is elaborated to delineate areas of acceptability for materials alternatives. This concept seeks to assist in making changes to the choice of materials for modifying an existing product.

The chapter on the impact of available production processes on design possibilities draws together some of the more important processes applied to materials, to remind the designer how an in-depth knowledge of these can affect his concept of a properly engineered and cost-effective product.

Summarising, it may be claimed that the product designer ideally needs a broad knowledge of the materials technology world, with its ramifications of materials sources, materials products supply, the behaviour of materials under different processing conditions, the factors involved in substituting one material for another, the question of preferred production methods for different product configurations, and many other aspects of the subject. These topics are addressed in the various chapters of this book to ensure that the person responsible for materials selection has an adequate understanding of the nature of specialist advice which he may and should seek, and that he has sufficient knowledge of materials to frame relevant questions and, most importantly, to understand the experts' answers, be he building a safety-pin or a road bridge.

Apologies and acknowledgements

For simplicity the term 'he, the designer' has been used throughout the book, being intended to apply to any individuals of either sex who, although not necessarily trained in materials-related disciplines, have responsibility for selecting materials used in a manufactured article.

After a lifetime of work with materials, this author, as would any other, finds it impossible specifically to acknowledge all the sources of detailed information which have been drawn upon. Much use has been made of data drawn from day-to-day reading, as shown in the list of References, and peer discussion, coupled with that built-in mental awareness of the main truths about materials which is the residuum of student days. Most of the references cited are to specific papers, and as such are recorded in the References list. Occasionally however, a text listed in the Bibliography is known to contain specific data under discussion: the author and year are cited as for references. So an author 'missing' from the References will be found listed in the Bibliography.

A special mention should be made of my colleague from the British National

Committee on Materials, the late Nigel C.W. Judd. It was he who provided the motivation for this book and I am indebted to his memory for the many fruitful discussions and interchanges of information which we had before his untimely death.

Thanks are also due to STL Ltd for providing considerable assistance, to Mrs J.R. Newcomb for her untiring work with the word processor, and to Mrs J.M. Trentham for producing the illustrations. Mr M.V. Coleman, Dr M.J. Folkes, Mr P. Green, Mr A.C. Harman, Mr F. Kerry, Mr P.J. Lesley, Dr K. Paul, Mr J.Pemberton, Mr S. Tattersall, Mr K. Taylor, and Mr V. White deserve the author's thanks for making valuable comments on various chapters.

Some of the concepts in the book, notably views of material flammability and of materials 'envelopes of properties', are considered to be novel, but doubtless, like almost everything else, they have been developed before, forgotten and then rediscovered. While acknowledging, in Figures, Tables and References, at least some published authorities, the author begs the reader's indulgence by admitting that countless scientists have shaped his thoughts, and so he thanks the world at large. Specifically, however, acknowledgements are due to E.I. DuPont de Nemours for permission to reproduce Figure 3.6, to the Institution of Production Engineers for permission to reproduce Figures 11.1, 11.2 and 11.3, and to Mr K Thomas of the Institution of Structural Engineers Materials and Components Study Group for the concept and much of the content of Table 3.4.

However, the author can claim something specifically for himself; that is, full responsibility for the absence of that information which should be included in an ideal book reflecting an ideal world. Such omissions are entirely his own responsibility and reflect that sad lack of omniscience and of available pages which so many authors share.

In parts of this book, mention is made of trade names, specific materials and proprietary processes. Their inclusion does not imply that the publishers or the author have tested them, and they thus cannot endorse what may well be worthy additions to the designer's armoury.

E.H. Cornish 1986.

Introduction

The need to know

Much of the world's wealth is the product of manufacturing industry. Even in instances where a valuable commodity is not specifically shaped into a manufactured product, as, for example, oil, minerals, diamonds or gold, nevertheless elements of engineering thought go into their processing and final presentation to a customer. However, the vast majority of manufactured goods represent the fruits of teamwork in industry, involving activity by planners as well as by machine operators. The artifacts with which we are all surrounded have been manufactured industrially on the basis of ideas from a variety of people. Being made of materials of one kind or another, it follows that the materials must have been selected by somebody. One purpose of this book is to help that somebody, be he a designer, a practising engineer, a student or whoever, to become more aware of the world of materials technology with which he has to be concerned.

In the twentieth century, no man can be an expert on all those aspects of a saleable article. Even though a craftsman whittling a wooden object may have a deep knowledge of the characteristics of wood as it responds to a sharpened carving tool, he will probably know little of the complex chemistry and biology of the material on which he is working. Even more remote is the likelihood that a designer, choosing a metal for a particular application, will be equally expert in its chemical corrosion properties, its morphology, its strength characteristics and the routes by which the metal is processed, to mention only a few of the aspects affecting selection. Let us assume, therefore, that anyone selecting materials will more often than not need the benefit of expert advice. One of the first questions to be raised, of course, is: 'Where can this advice be obtained?' This book seeks to answer that question and also many of the others which will crowd into the originator's brain at the beginning of a new project. How much needs to be known about service conditions to which the product will be exposed? What are the manufacturing routes by which it can be made most cheaply? What constraints on supply of the material might operate if quantities are required over a short or long period?

As well as aesthetic and technological problems posed by a new product concept, the designer has to fight other, more mundane, battles. He needs the support of his employer, who in turn might need convincing that the designer's contributions as an originator, and as a coordinator, are vital to the business. The designer himself has to be aware of the various factors in engineering, materials technology and production processing (to name only a few) of which he inevitably must be a focus. He has to ensure that the design brief given him is relevant, complete and timely. These aspects of the design activity are covered in the early part of the book, to set the scene for its main thrust, that of available materials properties and routines for their selection.

The product in its environment

At the outset of a design study, the constraints on the project and the requirements for the new product have to be assessed. Compared to preceding similar products new factors must be identified, such as possibilities for improved or wider ranging performance, the effects of artificial and natural environments, reliability expectations, aesthetic features and achievable economies in materials and energy during production, service and later recycling. Increasing demands are always being made on materials, requiring them to present new capabilities in extreme environments. These, coupled with the need for low lifecycle costs, force the designer to think in terms of improved reliability and of lower failure rates when compared with past performance.

Apart from having to withstand natural atmospheric weathering conditions, materials for aerospace, marine, chemical and industrial engineering applications will be subjected to various specific artificial environments. Temperatures experienced by a material may range from those attained momentarily on the surface of a space re-entry vehicle, to the continuous sub-zero temperatures imposed by space exposure or in modern cryogenic systems. Exposure of materials to ultra-violet or other high-energy radiation will depend not only on the immediate world climatic zone but also on distance from the earth's surface. Contact with water may range from the very low humidities at certain parts of the earth's crust to deep submergence situations in oceans. Additionally, while in service, materials may be subjected to long-term sustained stresses (potentially leading to creep) or to cyclic stresses of constant or varying amplitude, perhaps imposed by acceleration, vibration or impact, and likely to produce fatigue effects.

Consequently, during the design and development of a product, knowledge of the environmental performance of candidate materials is essential to ensure reliability in service during an adequate and predetermined lifetime.

The materials cycle

Materials may be described as one of the connective tissues of national and world economics; recently a system-concept has been derived; the 'materials cycle', of which the designer must be aware.

The materials cycle reflects the situation that only a few materials are extracted from the earth by mining, drilling and harvesting. They are processed into bulk supplies, for example of metals, cement, clay, timber, natural rubber and petrochemicals. Alloys, glass, ceramics, bricks, plastics, elastomers and composites are produced from these bulk materials and subsequently worked up and assembled into structures, machines and consumer products. After use, discarded products are returned to the materials cycle for reclamation or for disposal by burning or burial. Many sectors of the materials cycle show strong interactions between energy utilisation, materials supply and the environment. For example, careless disposal of rejected waste from metal extractive processes can produce undesirable social and biological effects on general and local environment, and can also lock up other metals in a form not economically recoverable.

The close interplay between utilisation of materials and energy is particularly evident in the US economy. About one third of the energy consumed by American industry relates to the value added to extracted primary materials by their purification, production and subsequent fabrication. Selection of materials suitable for releasing or converting forms of energy is itself crucial. Many of the advanced energy conversion technologies such as gas turbines, nuclear reactors, magneto-hydrodynamics, solar energy and fossil fuel conversion, are at present materials-limited in terms of efficiency, reliability, safety and cost effectiveness. Even renewable energy resources such as wind, waves and biomass pose secondary materials problems in construction of effective operational machinery.

The materials cycle provides an analytical framework for examining perturbations in materials availability with respect to the whole world, a country, an industry, a company or a workplace. Flow rate of materials anywhere in the cycle can be affected by economic, political or social decisions made in other parts of the circuit. When analysed in this context, materials shortages are often found not to be caused by actual scarcity of natural resources, but rather by dislocations in the cycle which interfere with the anticipated arrival of materials, at appropriate manufacturing stages, in the required amounts and at reasonable prices. An example of this occurs when processing or transportation capacity in the car industry become inadequate somewhere in the cycle; this is seen at times of industrial unrest when components suppliers are unable to deliver the goods and centralised car assembly is thus impeded.

The role of processing

Reliability of an engineering structure or a component depends on the materials used, their properties, the influence of uncorrected defects, the processing or fabrication technique selected (and especially on any sensitivity of the materials to processing), and response to service environments. Failure of a material or product under stress is determined by the growth of defects which may originally have been present, and the subsequent propagation of cracks. Reliability demands the application of manufacturing techniques which reduce the incidence of crack initiators.

Optimisation of processing conditions is not easy, and fabrication and service requirements may conflict. For example, incorrect fabrication of fibre-reinforced synthetic resin composites can give rise to thermal stresses which may induce crack formation during subsequent use. To minimise this effect, low temperature curing systems should be selected. This, however, imposes a constraint on subsequent high-temperature use in service, because resins suitable for such performance usually require an elevated temperature cycle to cure them properly. A compromise between optimum service performance, processability and cost benefits may be necessary. Because of this, materials engineering in the true sense of the words has to be employed to select a 'best material' for the intended end use.

Materials structures

Materials science, technology and engineering are interdisciplinary activities concerned with the generation and application of knowledge relating to the composition, structure and processing of materials, the properties of which control end uses. Some parts of this spectrum of activities, when observed narrowly, may be regarded as science-dominated, whereas other parts may be engineering-dominated. No person educated only in a single discipline can hope to excel unaided in this broad field.

Literature offers excellent descriptions of the history of materials technology and materials, and of their use in jewellery, other art forms, ceramics, textiles, tools and weapons. Invariably, attention is drawn to the transition from the craftsman, with his molecular composition materials approach, to the present day structural — atomistic approach adopted by the professional technologist.

The structural — atomistic approach to materials has largely derived from increasingly penetrating experimental techniques to investigate their internal structure. Optical microscopy can be used to examine coarse microstructure, electron microscopy interprets sub-structure, crystal and molecular structures are investigated by X-ray diffraction, and atomic structure by various excitation spectroscopy methods. Nuclear structure is elucidated by high energy bombardment of materials with sub-atomic particles.

Scientific knowledge accruing from these investigations has accelerated development of new man-made engineering materials such as synthetic fibres, plastics, elastomers, high strength and high temperature metal alloys, ceramics, glasses and composite materials. It behoves the designer to know something of these, and they are discussed in later chapters.

Engineering materials can be categorised either as crystalline solids, in which atoms and molecules are stacked in a more-or-less regular array, as with metals, or as amorphous materials having unique rheological properties and no long-range atomic structure. Plastics and elastomers fall into this class. There are, of course, many intermediate forms of structure: few amorphous materials have no crystalline areas at all; metals are always crystalline unless very specially treated.

The study of materials science, technology and engineering has been defined in terms of three different levels of materials structure, which control properties and from which an understanding of materials behaviour can be obtained.

At the engineering level, a material can be considered as continuous and homogeneous, average properties being assumed to apply throughout the whole volume. This approach rationalises evaluation of mechanical properties with large test specimens, under conditions closely resembling those prevailing in an engineering structure. One example is the use of steel to fabricate a beam.

A material at the microstructural level may be considered as a composite of different phases. These phases may be similar, as in metals, and may be in thermodynamic equilibrium, or may comprise a mixture of dissimilar and unrelated solids, for example fibre-reinforced composites. At the microstructural level generalised treatments for the understanding of multiphase materials behaviour can be developed. This approach can explain the mechanical behaviour of cast iron.

At the molecular level, a material is considered as made up of discrete particles of molecular size. This concept is used to explain the melting behaviour of solids. Although there is some limited relationship to the engineering and materials structural levels, the discrete atomic particles and discontinuities at molecular levels represent a different order of behaviour, and offer difficulties in visualising conceptual links with the other two levels. Thus one cannot readily look at a molecular structure and deduce from it a valid range of materials properties in samples of a size appropriate to engineering uses.

Stemming from an understanding of materials microstructural and molecular status, modification and development have led to advances in polymer molecular engineering to produce, for example, isotactic and syndiotactic polymers. Similarly, in metallurgy, developments include production of uni-directionally solidified alloys; glasses have been highly refined in the quest for communication by fibre optics.

The foregoing exposition seeks to convince the reader that more than a superficial review of published materials properties is needed to achieve proper comprehension of the subject.

Selecting materials

Selection of material for a specified application is one of the earliest and most important design decisions in product development. Expert judgement is required to ensure that the material will be suitable with respect to cost, for manufacturability of the structure or component, and to meet in-service performance specifications. Such considerations are not mutually exclusive, but all decisions about them must be recognised as compromises balancing trade-offs between different levels of reliability, performance and maintainability in order to achieve minimum lifetime cost.

For good design decisions experience is irreplaceable, as it is seldom possible to analyse in depth, from first principles, all aspects of an engineering problem. As many decisions may perforce be made using inadequate data, it is obvious that seeking additional expert help should be considered in all but trivial cases, in order to broaden the experience base available (and to share the responsibility!)

Designers are ultimately limited by the properties of materials which can be obtained commercially. To make effective use of available data on materials properties, they should be aware that reported property values are influenced by variations in composition, by processing technology and by the test methods used to obtain the data. A 'gut-feeling' view of materials science will lead to appreciation of how property limitations might depend on the structure of material as considered at the three levels described earlier.

1

The impact of design on manufacturing industry

The first assertion in the Introduction was that the designer has a key part to play in creation of wealth by manufacturing industry. A little more exploration of this theme seems justified to deal with the role of the design function and of the designer personally in industries which create goods from lower orders of materials supply. Various parts of this chapter will review how industry in turn sees the way in which design can serve its needs, and will focus on those specific contributions which the design office and the designer himself can make, provided that sufficient support facilities are offered to him. The chapter is rounded off in moving towards the central theme of this book by reviewing those aspects of design which are specifically important when selecting materials.

1.1   The role of design in product manufacturing

The prime purpose of design in industry is to maximise the value which purchasers of manufactured goods are likely to place on the finished product. Naturally, enhancement of this value is what manufacturing activity is primarily concerned with in a capitalist system. Industrial design recognises well-established customer preferences in the market place, often defined in terms of colour, texture, shape, safety features, and physical proportions, as well as less tangible manifestations of 'joy of ownership', such as reduction or elimination of dirt, heat, smell or noise, which might distinguish the product from competing or earlier offerings of the same kind. Furthermore customer preferences can sometimes, at least in the case of ephemeral products, be focused on novelty or on innovation for its own sake. Of perhaps greater importance to the controllers of industry is the thought that design can enhance the value of a product, in relation to competing products, by providing increased performance, durability, strength, improved functions, greater versatility or, of course, less cost.

To achieve these ends the design function in a company or industry has to apply what are, by now, well-proven principles to ensure that all new features are properly evaluated and are then adopted, modified or discarded in accordance with the market being addressed. In this way, the manufacturing company can provide what the customer wishes to possess and is prepared to pay for.

Even in the present period of economic stringency and slow emergence from a long period of recession, still not enough people in industry or in the market-place recognise the importance of good design to the success of a commercial enterprise. Although in the marketing of any goods, timeliness of delivery, quality and reliability must be maximised to compete successfully in world markets, particularly in the engineering industries, the other factor, that of design, can operate by making the product more appealing to the customer and hence provide a market-place advantage. Design also plays its part within the parent industry or company by improving the ease of production and the productivity of existing plant, thus enhancing the creation of wealth and retention of jobs.

That the design process must continually operate even in an established manufacturing operation, is obvious from a consideration of Figure 1.1, Judd (1984a). Here we see that each product offered in the market-place has a life-cycle. Research and development (R&D) enables its introduction to be effected, prior to the period of growth during which the product finds acceptance. After a while, it becomes mature, either through built-in obsolescence or as a result of new developments; by this time the far-seeing company will have replacement products already in the R&D stage. Inevitably (and this may occupy a period of months or of decades), the product will go into market decline, and decisions must be made as to whether any of the design features can be retained to produce a new revitalised product, or whether the operation has to be closed down to make way for an entirely new family of products.

Leslie (1982) discusses clearly the impact of industrial design awareness on a large multinational corporation. As new products become increasingly complex, and as more functions are required, improvements in industrial design and revision of human factors engineering become necessary. These together ensure that complex equipment can be accepted by a wide range of operators to reduce the likelihood both of long-term operating stress and a significant error rate in use. Another feature of the design process in industry is to enhance 'corporate image' which, of course, is of real value to the originating company only as long as its goods retain a well-deserved good name in the market-place. Leslie shows the necessity of a management programme at sufficiently high level to enable an industrial design department to be set up in the company. Not the least part of such a programme is recognition of the need for liaison between designers and experts in other fields such as materials, surface finishes, manufacturing technology, associated component part technology and international standards, some of which themes will be explored in later chapters. An industrial design team has to operate in close conjunction with other departments in a manufacturing enterprise; it is, as a consequence, very important that these other departments see their interaction with individual designers as an essential feature of success in the business, and not just as an irritating and unproductive side-line. This puts the onus back on designers, as they have to be aware of the basic knowledge available to the other people with whom they have to deal, so that they can discuss problems intelligently and, even more importantly, can understand the answers they receive from these other experts.

There is still a belief in industry that an industrial designer is no more than a stylist who attractively packages a product and, if lucky, provides some sort of a corporate identity to a range of related products. However, the true industrial designer has much more to offer than this. He must be able to appreciate all aspects of the service operation of equipment, to consider form and function independently and, by his creative skills, visualise a concept at an early stage and show it in a workable form which will appeal in the market-place. Technically sophisticated products may often only have been operated as laboratory models, and the designer has to transfer this functional concept into an attractive and useful product, often by considerable redesign of the original idea. The impact of a designer is, of course, maximised if he is brought in at the very earliest stage of a product concept, and this implies management awareness to

Fig. 1.1. Life cycle of a product. (After Judd 1984.)

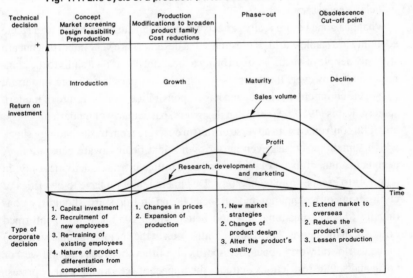

ensure that he is involved in meetings with others whose expertise will also be brought to bear on the problem.

Another area in which industrial design, and indeed the designer himself, can assist in maximising company profits, is to be aware of alternatives to traditional constructional materials, methods, surface finishes and so on, and perhaps also to be aware of newer production processes which might, if adopted, revolutionise the engineering and cost concept of a new development.

In short, although the design process varies between products, there are usually several well-defined stages in any development which should be followed in an accepted order. The first important step is to carefully define the brief given to the industrial designer, in order to establish the objectives and the various stages which are necessary to achieve them. Figures reflecting anticipated production quantities, and certainly the target selling price, should be produced at an early stage, since these can significantly affect the chosen design, which, of course, is always influenced by the existing manufacturing techniques and availability of plant at the time.

The second stage of the process is to look at the product concept as though one were a user, in order to determine how the product would be handled, by whom, and when. At this stage there could be involvement with codes of practice, accepted trade practices, patents, safety aspects, standardisation, and so on. Only then should the designer produce a number of sketches or models to illustrate his ideas, having collaborated with other technical experts so that their united skills can be brought to bear on the problem of achieving maximum innovation and marketability. Having done all this, the designer then finishes off the project by supervising the production of specifications, working drawings and models as necessary.

What has gone before really reflects the optimum situation based on a stable economy in a stable world. It presumes that designers are very important but are still only people in a team who can bring to bear only their own skills to the common good. However, there are views that the designer has a more important role in the community at large, and has responsibilities for the future of modern society. Rams (1983a) asserts that designers, particularly of modern products, are often confronted with adverse criticism deriving from the technology shock which sometimes occurs even in our own scientifically aware generation. A designer cannot drop out of society and take refuge in artistic activities only, in an effort to escape from technology. Designers really do have to be critics of civilisation, of technology and of society, and in this spirit should try continually to find something new and better which can stand up to informed criticism in the future. The designer must look at the socio-economic structure of the world and worry about socio-political conflicts between east and west or north and south, restrictions in trade, the impact of see-sawing energy prices, the effects of bureaucracy in controlling imports (and much else), the scarcity of raw materials and, not least, the inertia inherent in a massive industrial system. From this point of view, he has to be well informed about what is going on in the world, without waiting for a government commission to set up a survey and tell him a year after the problem has become serious. Quality, simplicity and integrity in design are Rams' touch-stones for the industrial designer of the future.

1.2   The designer's contribution to the requirements of industry

In order to succeed, any commercial company must recognise the status of the market-place in influencing the timing, and design and price of products to be manufactured. The goods have to meet all relevant market criteria, and their specification must also be a good fit with the ultimate customer requirements. A continuous review of market requirements is needed, not only at the initiation of development, but right through product specification and all the other stages, to the point of freezing the design for initial manufacture. This ensures that the end result is what was originally planned, and not something which turns out to be more expensive, less reliable, harder to make or otherwise reflecting some tenacious misjudgement on the part of the development team. Many potentially fatal errors can, of course, be reduced by adhering to regular design review procedures through meetings involving all the appropriate company functional staff.

In a competitive situation (and often within and between R & D departments in the same company), it is occasionally difficult to discern the various pressures which motivate the creation or development of new products. One pressure might be top management's need to improve corporate productivity, and another might be a self-interested thrust from development and sales departments who effectively can invent a need. Judd (1984*b*) reports examples of significant innovation within 116 organisations, classified in Table 1.1. Over half the innovations were ascribed by the organisations concerned to causes

Table 1.1. *Reasons given as the driving force for significant product innovation*

Reason for innovation	Number of cases
New product to extend normal range	105
Improvement of existing product	13
Use of new processes to assist production rationalisation	16
Use of new processes to expand production	8
Changes in production methods	62

such as the pressure of competition or of excessive demand on a restricted manufacturing capability.

It has been said by Bishop (1982) that most designers have a lot to learn. The point Bishop makes is that for a designer to initiate a really successful design, he should be capable of using the product in question. It is generally accepted, for example, that most motor mowers are rather better designed than many medical instruments, from the users' viewpoint. A designer of word processors is presumably more effective if he is able to type. Although in many cases, of course, this precept cannot be followed literally, it brings out the point that designers of products should essentially be practical people and not merely high-grade draughtsmen or artists.

This brings us to the consideration of what top management of a manufacturing business might expect from the in-house function of industrial design. Any thinking designer will have his own ideas regarding his top management's expectations of the design department but this, of course, is a viewpoint which might tend to become parochial. Graham (1981) suggests that industrial designers should not attempt to educate their top management on the concerns of design, but should put more effort into enquiring how management thinks design can improve profitability. Management should ask the industrial designer to ensure that the outside of a product reflects what is inside. In these high-technology days, the highest technology in electronics or mechanical engineering may be hidden inside a plain outer case. The customer, at a cursory glance, sees no more than might be offered by a lower-technology competitor. Top management perceives the problem that visible parts of the product should reflect the quality and level of technology of the internals, and relies upon its designers to achieve this miracle.

However, in reviewing externals, management will thus be asking the industrial designer to act as an authority on matters of taste in visual elements such as colour, form and texture. To retain credibility, the industrial designer should have a good track record of past performance and sufficient reputation to inspire justifiable confidence. He should be expected to explain the position of his product relative to its competition, in design reviews and in other in-house discussions, and also should be capable of reflecting the character of his company in the product's design.

The industrial design department of a company is also expected to protect the investment made by management in the long term good will of the customers. Ideally a company will have a well-respected name and easily identified trademarks, a reputation for quality, for fair warranty service, and for ready availability of its goods. The executives of such a company zealously guard this invisible asset and demand design decisions to bolster the long-term trend.

Industrial designers must be aware of this view when putting forward any avant-garde ideas. Lastly, and perhaps most important in the difficult and intellectually demanding position in which the industrial designer is placed by his top management, is the question of his being an effective negotiator during the product development process. The corporate chief executive is an orchestrator of business functions. He is aware that each of the many departments in his company must make critical contributions to the development of products, and is also aware that there are conflicting motivations and values in the various Directorates. To overcome potential conflicts, the management teamwork aspect is stressed, and there must be a balanced contribution between departments, no single one being left out or allowed to dominate the others. If the industrial designer is inflexible in what he will accept as an assignment he becomes an obstruction to successful product development and cannot be tolerated. On the other hand, he must be an effective force in protecting the management expectations of the design function as product development proceeds. Rams (1983b) gives examples of expectations of the design function in contributing to company success. He asserts that there are yardsticks for the quality of design which can be applied to any type of product:

First of all, there are the basic requirements of function and performance: every article, except works of art, has to fulfil some sort of function and it must be designed in such a way that the various requirements for its use are best met - the more intensive and demanding the use, the clearer are the requirements. As a second requirement, the design must be feasible, which means that within constraints of budget, the present state of the art, time for development and other factors, the optimum design result can be achieved; of course, aesthetics are always a requirement, as man has need of beautiful things. For an engineering object, 'beautiful' may perhaps be too extreme an adjective, but at least the designer should avoid a product which is ugly, even if the beauty arises only from a properly chosen decorative finish. Top management cannot be provided with all the possible designs for a product, but have to be guided towards a small range of well-considered alternatives. Chief executives cannot by industrial protocol go to a drawing board or build models to show what they want. They can, at best, only accept, reject or influence a change; the impulse must derive from the designer.

This brings us then to consider the status of design and of designers. Design should be a creative, exacting and responsible occupation. Successful discharge of the designers' duties should improve their status within the engineering profession, to a level which their responsibility deserves. Management should ensure that the design function in their companies is supported by incoming young people who are to be convinced that design is a worthwhile career, and that the function itself should operate with a sufficiently high level of command so as to ensure that its inputs to a new product development programme are considered seriously.

1.3   Management functions associated with design.

Product design and management are often not sufficiently accepted by top industrialists as key functions in their business. The Council for National Academic Awards has published a report (Oakley 1984) advising managers on how to achieve greater awareness of, and interaction with, the designers' needs. In an ideal world, design should specifically be identified as a Board level responsibility, on a par with production, finance and R & D. Without such visibility, it cannot be expected that the full potential of successful design would be realised at minimum cost. Possibly companies should designate a Board member to take care of the design function as part of his responsibility. In particular, this member should ensure that marketing, production and finance groups are sufficiently involved when design decisions are made, and that, further down the line, departments such as purchasing are also kept in the picture.

Another Board activity is investment in new plant such as, for example, computer facilities. Too often, at present, investment decisions exclude the design function, although the effort involved in designing a product is as much an 'investment' as is the R & D leading to its realisation. Apart from the the obvious investment possibility of providing computer-aided design facilities in both hardware and software, investment of time and effort by the business should include a product plan which is used as part of the design review process mentioned before. Table 1.2 lists some questions which (again in a perfect world) might be posed by top management to their design team.

There is always a risk that product design may originate in isolation from the activities of other key business functions. The result of this is that, at too late a

Table 1.2. *Business criteria to be satisfied early in a design*

---

1. What is the intended market?
2. Is there a balance between selling price and expected product life revenue, bearing in mind the start-up costs?
3. Is manufacture going to be economic?
4. Is up-to-date technology incorporated in the design?
5. Are materials and components available?
6. Have reliability and maintainability aspects been considered?
7. Have the critical phases of development been properly planned?
8. Can market launch be achieved at the desired date?

---

stage in the development, the design has to be adapted to meet all kinds of requirements not adequately identified at the outset, and to accommodate unconsidered limitations on production processes available, as well as to accept cost budgets imposed by other people. These are dangers of hamstringing the design department which good management will avoid.

1.4   Design activity in a company

The industrial designer's job is to develop the design of three-dimensional products, which may range from very cheap and simple consumer goods to more sophisticated equipment such as machine tools, optical or electronic equipment, or even more complex devices such as vehicles. His interest is to make the product work, not only in terms of achieving a sale in the market-place, but subsequently in providing satisfaction to the owner in the areas of cost, maintainability, ergonomics, safety and convenience. The designer must be competent in basic design skills, the first of which is problem identification and analysis, followed by ability in both free-hand and engineering drawing, coupled with an artistic understanding and sensitivity towards colour and form. Additionally, he must have, at least within the context of his own employment, a working knowledge of engineering technology and, to some extent, of the principles of marketing.

Whether working alone or in a team, a designer has a sequence of operations to progress, which must fit within the overall management organisation of his employer, whether this be an industrial company or a short-term client. These activities include agreeing a brief which defines the extent of his involvement, establishes objectives, identifies with timed milestones and the various development stages of the product. Then he has to take cognisance of how product requirements are affected by the production methods available, ultimate costs, methods of distribution and marketing. The designer must then satisfy himself that he can effectively identify the future user's expectations of the product and should follow this by obtaining information about current legislation, safety requirements, standards and surveys to ensure that his concepts are going to be legally viable. Only then follows the more popular view of the activity, involving production of sketches or models, and eventual development of specifications and drawings. These activities are illustrated in Figure 1.2.

Although most of these requirements seem obvious, quite frequently industrial companies, through lack of management recognition or sheer unavailability, do not commit adequate resources to some of the most important aspects of product design. These include developing the initial specification requirements and making some measure of the cost-effectiveness of the design proposal. More effort is often needed in reviewing the development's progress through various stages, in particular with respect to holding properly organised design reviews in which the designer's activity can be considered side-by-side with technical, production and other functions.

A design department might find itself embedded in all manner of organisational structures within a company, ranging from the simple 'pyramid' or 'multipyramid' types of organisation to the more modern matrix organisations in which it interacts with engineering management, mechanical engineering, the test department, and also with research, new projects and individual project

Fig. 1.2. The designer as a product generator. (Copyright: 1985 Standard Telephones and Cables plc.)

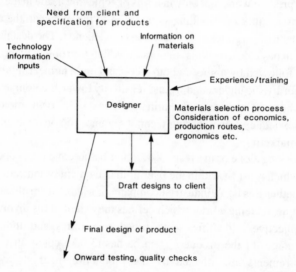

Fig. 1.3. Relationship of design to other activities in a company. (After Judd 1984. Copyright: 1985 Standard Telephones and Cables plc.)

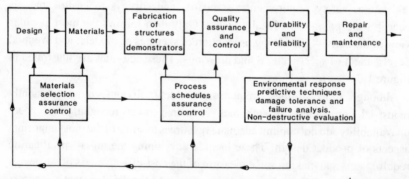

designers, to offer just one example. In any event, the designer should be familiar with many other activities within his company, as shown in Figure 1.3.

An internal industrial design activity usually arises from small beginnings, seldom coming about through a positive act by management to set up a full team at the beginning of a Company's life. However, an expanding small company should at least employ a competent design manager and invest him with full authority and responsibility for carrying out industrial design activities; he needs accessibility to the top executive. Depending upon the situation in the company, he can then rely entirely upon outside consultants, create an exclusively internal design department, or use a balanced mix of both. In practice it would seem that a small company should entirely depend upon outside consultants, although this does not absolve it from the need for an internal expert who has charge of all the projects, and can act as artistic arbiter and financial controller of the activity. A larger company may gradually build up some in-house capability in the design department, but even a very large company may occasionally need outside consultants, although in this case they will, from their own experience, be very well able to brief that consultant adequately from a position of knowledge. Consultants should be called when specialised expertise is needed to solve a particular design problem. In accordance with the theme of this book, it must be pointed out that the design manager or coordinator must enjoy continuous liaison with experts on materials and surface finishing, as well as on manufacturing technology, the technology of sub-assemblies and components, and international standards; some skills in ergonomics, human factors and marketing should also be available.

1.5   The designer as a focus of knowledge

Having described in the previous section the requirements for a paragon of the virtues who knows everything about design and about most other aspects of the business in which he finds himself, one may wonder how this design coordinator or manager becomes adequately educated. The designer needs basic information on new technology, new manufacturing techniques, new materials and components, and depends of course on both formal and informal methods of learning, especially for subsequent updating. Later in this book we shall see how data banks and other information retrieval sources are available, but we must not underplay the importance of less formal contacts such as attendance at specialised conferences, and contacts with trade, industry, universities and government laboratories who have relevant viewpoints to deploy on request. Research and trade associations are also important for transferring technology about specific subjects to the interested enquirer. Figure 1.4 highlights the designer as the focus of a web of information whose purveyors he must identify and cultivate if his job is to be carried out well.

The designer has to be adept in extracting a proper brief from those to whom he is ultimately responsible for his contribution to a new project. The brief should embody all the information necessary for him to perform the work he has to do; background information should be available to enable him to reach fully informed decisions, at least in the initial stages of the design process, at a time when he may need to work very much on his own, without necessarily having too much contact with R & D or other departments in his own organisation. Inputs to the brief would come from marketing, engineering and other departments which should be prepared on demand to put full information into the project at this early stage. A useful concept is to develop a checklist to elicit as much information as possible, and to have this circulated and completed by all appropriate departments before a main briefing meeting so that much unproductive argument is eliminated from the start.

A designer is ever at the centre of activity involving transfer of technology and research results into his memory bank. He has to visualise the available materials and components, and the production technology and manufacturing

Fig. 1.4. Management functions and expertise associated with the design process. (Copyright: 1985 Standard Telephones and Cables plc.)

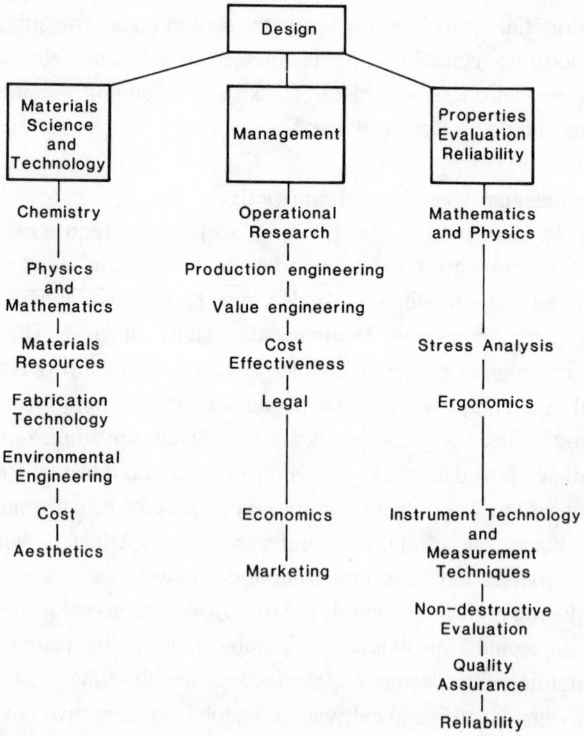

techniques available in his company and in his industry. He is essentially dependent on knowledge of the latest state-of-the-art in his area of work. A vital function of his company management is to ensure that facilities for obtaining this information are available from internal library sources or from people (known in the Massachusetts Institute of Technology as 'gate-keepers') who take care to be abreast of new information and circulate it to interested parties. A designer is seldom able to read much of the specialised technical literature in any depth. In fact, he may have difficulty in reading through even those journals devoted specifically to design. Much better than fragmented browsing through books would be to arrange for a supply, on a regular weekly basis, of information from the Selected Dissemination of Information Services. In these, a number of keywords are identified and the operator, via a computer link, is able to interrogate appropriate sectors of the world's open literature. Unfortunately, few people are good at specifying apposite keywords, so they cannot be sure that their interests are properly covered by this system unless an experienced librarian/terminal operator is available to them.

The subject of educating design engineers in materials technology has recently been reviewed by Miller (1980). Supposing that designers of products can never have too much information made available to them on the capabilities of materials technology, Miller focuses attention on the respective roles and responsibilities of the metallurgist, the mechanical engineer and the industrial designer, and discusses how the gulfs between them might best be bridged.

1.6   Attributes of good product design

A purist would identify many aspects of design which must simultaneously be considered and kept in balance. A basic requirement is that the product to be designed should first be specified, and this introduces the concept that a design must cater for all aspects of the required performance. Cognisance must, of course, be taken of the manufacturability of the projected design options, not only in the absolute terms that it can be manufactured at all, but in the more restricted choices posed by the existing manufacturing resources of the company producing that product.

Turning then to aspects of design of importance to the customer, one is concerned with marketability and the stature of the product compared with its direct competitors, with the appearance, and with the probability of delivering the required quantities at the specified dates. Not least important are aspects of reliability in service and of maintainability for products which have not been designed for an ephemeral service life. The trade-offs between high reliability, ease of maintenance and low first cost, are formalised at the specification stage but probably decided by marketing forces rather than by technical or design staff. Furthermore, the anticipated product life is important, as it can be too long. A long life article may not only be initially expensive, but its presence at the user's premises may be a continual uncomfortable reminder that he purchased an article which increasingly recalls a byegone technology.

Good product design depends upon existence of well-drafted standards. Some specify attributes of candidate materials for the construction; others cover design aspects such as thicknesses of sections and permissible radii of stiffening webs; while others are related to safety and might also emphasise performance and functional requirements.

Perhaps the three attributes most important in the context of selecting materials are: function of the article; appearance; and ease of production. These must be in sensible balance in relation to the product's use. A highly technical device with a requirement for a high safety factor, such as parts of a nuclear reactor, will be designed for absolute reliability without regard to ease of production, and certainly without regard to appearance. Furthermore, ease of maintenance may be sacrificed in some cases; the domestic consumer goods market is replete with these. On the other hand, an item of street furniture, which has to be both functional and compatible with its surroundings, will have some attention paid to each of the factors of appearance, surface finish, cost of construction and maintenance requirements. A simple domestic appliance may be designed almost exclusively for attractiveness of appearance as a trigger for impulse buying, without regard for any more than a nominal service life.

Selection of materials, and particularly of surface finishes, has a prime effect upon product aspect and marketability. After all, even with highly technological equipment, the purchaser will, first of all, see only the outside casing, and ultimately only the eventual (perhaps different) user and maintenance engineer will make closer contact with the internal workings. Accordingly, great emphasis should be placed on surface finish, both for aesthetic purposes and for functional protection of the underlying constructional material. It is for this reason that a later chapter is devoted to surfaces finishes and coatings.

Having said that, many materials, of course, have their surface finish built-in. A familiar example is acrylonitrile—butadiene—styrene (ABS) copolymer, regularly used for the bodies of telephone subscriber sets and for many domestic appliance housings. This material can be moulded with a wide variety of appealing surface finishes, which considerably enhance the appearance of the product and, in some cases, may be vital to its function.

Of course selection of materials for a design reflects the ultimate selling price which must be achieved, so that, again, a balance has to be struck between selecting a material that is technically or aesthetically the most suitable and one offering the possibility of economic manufacture.

Materials selected properly can contribute to neat and clean design and styling by exploiting properties which eliminate assembly operations in a product which would normally be made of many separate parts. For example, a single plastic moulding may embody the functions of several parts which, in an earlier and more conventional engineering age, might separately have been made of die-cast metal and subsequently joined together. A familiar example of part reduction in everyday life, is the appearance of modernday aeroplanes, at whose self-supporting wings one looks in vain for the wires and struts which used to decorate aircraft of a bygone age.

A last example of design factors which impinge upon materials selection is the range of densities of materials, which can have an impact on the final product. Articles meant to be portable, for example pocket calculators or portable television sets, tend to be manufactured of light-weight but strong materials, whereas larger equipments, such as free-standing filing cabinets, are more often made of a material of higher strength and functionality, for which weight is a secondary factor. In any event, the designer will ensure that the basic strengths of his materials are not exceeded by the overall weight of components of a design, and that sectional areas are thick enough for the purpose intended.

1.7   Materials as a factor in design

In a commercial enterprise, the design department must always be aware of the availability of materials and components, of traditional types as well as those derived from modern technology. The days are no longer with us, unfortunately, when small users of materials could influence materials suppliers, on a short timescale, to develop special materials needed for new products. The stringencies of recession, leading to streamlining of the materials supply industry, have resulted in a reduced palette of materials available in any one country, so that designs are tending to stabilise on standard material offerings, such as sheet steels of specific types and of standardised section thicknesses. Less reliance can now be placed on obtaining every required material from the designer's own country, but the process of purchase from anywhere in the world is being simplified all the time, particularly by the efforts of multinational suppliers with networks of franchises and agencies. The onus is on the design department to maintain its awareness of the state of materials technology by continual reference to suppliers' information and to the technical and scientific press.

Competition between materials suppliers in different fields of endeavour is, however, a continuing factor which works to the advantage of the designer in manufacturing industry. Advances in technology favour first one, then another, of the spectrum of materials which might compete for specific applications. Thus, in the last century cast iron was very widely used for ornamental and constructional purposes. This gave way to the use of light alloy die castings, at least for small components, and these in turn have been giving ground to plastic materials which are easily moulded. However, the ever-widening scope of environmental requirements imposed upon new designs often means that plastic materials are deficient through lack of high-temperature performance, which in turn puts the emphasis back on metal parts. In their turn, plastic material suppliers respond by improving the performance of their wares through chemical changes, or by changes in compounding or incorporation of additives. The battle goes on between materials suppliers, so that the designer must continually be sufficiently aware of the situation to take advantage of the best materials available at the time for his purpose. This book is intended to focus his ideas on a particular point in time, in the 1980s, but does not absolve the designer from vigilance in the future.

Finally, the reader's attention is drawn to the option of achieving maximum simplicity in design as a means of best using available materials. Although in a wide range of products, the designer's responsibility is to design-in superior performance, quality and reliability, he should always be prepared to design-out material content, labour, production costs and overheads. These desirable objectives are best achieved by simplifying and refining continually what may start as a complicated concept. In order to design for high productivity and also to achieve high quality and controllable material cost, the designer should again consider the simplification and size reduction possible in his product, and the maximum reduction possible in the number of component parts. He should be aware of the capabilities, particularly of new materials, of performing multifunctional roles in a product (as in the case of complex plastic mouldings), which can often go hand-in-hand with minimum material removal operations. An example here is the possible use of a net-shaped compacted metal powder part to replace one otherwise manufactured by machining from solid stock, with the consequent wastage of swarf. Finally, the designer can play his part in simplifying the product and reducing its costs by aiming for standardisation of parts and reducing their variety as far as possible.

2

Expertise required for the design process

It would be impertinent for the writer, who is after all a materials technologist and not a designer, to attempt to specify, even in general terms, the whole range of skills which a successful designer needs to employ in his profession. However, the hypothesis is still held that a designer is likely to focus his skills primarily upon the mechanical, aesthetic and functional requirements of products, and would not be expected to possess a general knowledge of all other aspects of the design process which would be available to him in a perfect world. When arriving at design decisions, experience makes a vital contribution, as it must seldom be possible to analyse in depth all aspects of an engineering problem by reducing every parameter to a numerical value. In our real world these decisions must be supported by an intuitive feel for their consequences, being in many cases arrived at from inadequate supporting data.

The following sections emphasise those aspects of the designer's work which are closely linked to the process of materials selection. The catalogueing of unfamiliar topics throughout this chapter should remind the reader of his need for expert advice from time to time; here, as in the rest of the book, the content of this chapter is designed to make the reader sufficiently familiar with the materials aspects of his problem that he can formulate questions properly and have some hope of understanding the expert's response.

2.1   The importance of mechanical design

The designer of even a simple mechanism must deploy a wide range of information, even if unconsciously. For example, he needs a knowledge of ergonomics, safety, product reliability, and at least some understanding of the ways in which a chosen material must be shaped for a design to be compatible with available production processes.

Pope (1980) raises interesting speculations that costs may be built into a product by default, unless optimising design to the available manufacturing process is properly effected. For example, reaming might unnecessarily be specified, when boring would be sufficient. A sawn finish might be adequate for a link end rather than a milled one, provided that stress-raisers are not too prominent. In the case of a cylindrical wear-resistant surface, is the usual manufacturing sequence of turn−harden−grind really needed, or could a finally applied coating provide adequate performance? The designer's detailed drawings must convey to the production engineer those freedoms for cost saving which can be built into the product.

Mechanical aspects of a design may be based intuitively, and thus probably incorrectly, upon the maxim that a published material performance parameter is adequately represented by a single unvarying figure. However, Haugen (1982a) points out that several materials properties are in fact random variables, and for an accurate design several different values must be used with the application of statistical principles. For example, 'yield point' is a stress value above which strain increases rapidly without a corresponding increase in stress, and below which elastic behaviour would be expected. Many materials have a yield point above which they pass from a predominantly elastic condition to a predominantly plastic condition, but note the term 'predominantly'; there may be no sharp transition. 'Fracture toughness' is the minimum level of stress capable of initiating unstable crack propagation, but its measured value depends on the quality of the testpiece. These two examples taken together with similar considerations for tensile strength, elastic limit, shear strength, torsional strength, fatigue strength and ductility, should all predispose the designer to take care when using numerical data, and not to design to the limits of the material without good reason.

2.2   Cost-effective materials and manufacturing processes

The selection of a material with which to make a product is one of the earliest and most important design decisions. Expert judgement is necessary to ensure that a chosen material will not only provide adequate in-service performance, but will be acceptable to the producer, in the sense of combining the right price with the prospect of being fabricated economically. During the selection process, first cost of a material can sometimes seem to be impossibly high, leaving a question mark over the project. Some materials may appear incredibly expensive per unit volume, but will perform in a unique way for long periods of time without the need for maintenance. For example, a small high-temperature bearing deeply buried in an aircraft jet engine may be worth its weight in gold if it significantly reduces the need for dismantling to replace a cheaper, worn-out component. The cost of ownership, affected by materials choice, includes elements of the energy costs involved in producing the material and in fabricating it, as well as in operating the final product.

When selecting a material, there must be an economic compromise between achieving a required technical function and minimising the integrated product cost throughout its lifetime. The total cost ratio, or in other words money spent per unit of time, is equal to the sum of the development, material, manufacturing and inspection costs, divided by the lifetime period, to which are added the costs of operating and of maintaining the product after manufacture. It may not always be possible to assign definite figures to these parameters, but the designer should be prepared for them to be invoked in a properly costed operation.

The designer should be broadly aware of those manufacturing processes available to him and the way in which they may have to be varied to handle the material under consideration. General processes for forming, joining and surface finishing any material likely to be under consideration should be broadly familiar, even though the materials may range from ferrous or non-ferrous metals, refractories, ceramics and glass, through plastics and elastomers.

Whatever the decisions taken, the designer should always be alive to his prime responsibility for the project, and to the difficulties which could arise if materials selection is left to some indeterminate operator too far down the line. Doane (1984) describes the sad tale of the two-year delay in completion of the New York Exposition and Convention Centre. In this building, the space-frame construction, intended to enclose a four (city) block structure, required 18 000 nodes to connect 77 000 tubes of the frame. However, the architects left the materials details to others, specifically the contractors. The construction manager of the project let several contracts for the work, and the space-frame sub-contractors ordered cast nodes from a variety of foundries. This meant that responsibility for casting quality was many steps removed from the original designer and certainly from the ultimate owner. Faults which became apparent led to a very high node reject rate and took the project back to square one. Our reader should be warned!

2.3 The challenge of materials selection

Modern uses for products increasingly take them into more hostile environments, with a consequently greater challenge to the designer's knowledge of materials. Although environments can be categorised generally, as discussed in Chapter 8, in reality for any product the environmental envelope in which it has to perform can seem daunting (Table 2.1). So, at the start of a new design study, the requirements must be assessed in detail; they may involve exposure of the product to a hitherto unfamiliar environment, or it may be essential to achieve extended capabilities or improved performance. Of course, in the mundane world of normal industrial and consumer products, the main reason for starting a new design project is to maintain or increase sales or

Table 2.1. *Some typical environmental conditions for aerospace structures and materials*

		Transport aircraft	Helicopters (blades)	Tactical missiles	Satellites	Manned reusable spacecraft
Overall required lifetime			15 years		7 years	Some years
Environment under storage conditions	Temperature		−55 °C to 70 °C		constant carefully controlled	
	Humidity		up to 100%		constant carefully controlled	
	Mechanical	negligible	10^7 Pa; Gust 3.10^8 Pa		negligible	
	Miscellaneous	in contact with water oil, fuel, microorganism		vibrations, shocks		
Operational lifetime		45 000 hr and 30 000 flights	5000 hr	a few minutes	7 years	Some years
Operational conditions	Temperature	from −55 °C to +50 °C	−49 °C to +50 °C		cycles from −150 °C to +150 °C	−150 °C +150 °C
	Humidity		up to 100%			0%
	Mechanical	vibrations 140 dB 30 000 cycles		vibrations 10 Hz to 2000 Hz	30 to 2000 Hz under 3 to 15 g (launching)	
	Miscellaneous	UV radiation, rain hail and dust			vacuum, radiation	

market share for an industrial company, and so considerations such as cost, aesthetic features, energy conservation and economics of use are probably more important. Costs may intrude in a number of ways, ranging from first cost of the material through the cost of manufacturing to the cost of applying an attractive surface appearance to increase customer acceptance. Duckworth (1984) eloquently promotes the thesis that the hitherto substantial increase in quality and variety of materials of construction is now over, and the challenge for future engineering generations is not to develop additional materials but to realise more cost-effective methods of manufacture for those we already have.

2.4   Materials aspects of design
We now arrive at some specifics of materials selection.

2.4.1   *Stages in materials selection*
Selecting the proper material is the key step in the design process and is the decision linking on-paper numerical calculations and engineering possibilities on paper with a real model or practicable design. The decision process must rely very largely upon the designer's previous experience and a knowledge of what materials are available to him; it must be borne in mind that there are at present over 40 000 metal alloys which could be selected and many thousands of non-metallic engineering materials. Selecting the best material for a part involves not only its price, but the cost and method of processing it to the required shape. At the same time the material must be able to withstand the environment to be imposed upon it. Dieter (1983) points out that until quite recently materials selection was often done from handbooks listing properties, but the choice was then constrained. In advanced aerospace and energy generation applications, materials are expected to operate near their limits, so specialist data has to be available from materials suppliers. For products subject to much customer reaction, for example in the automotive field, the driving forces change to reduced weight and minimal first cost, requiring different sets of data. In these days of suppliers' rationalisation programmes, one difficulty is that materials shortages may be actual or likely, so that selection of materials on a historical basis might not always be possible owing to non-availability.

Early on in the selection of materials, assuming that the requirements are well-defined, a screening process must take place to narrow down the possibilities, supported by published design data. These are fitted to the temporarily chosen material to approximate the sectional thicknesses and geometries which can be chosen, and verification of satisfactory operation should be based on laboratory tests or computer simulation. This simple sounding sequence can become very complex in high-technology applications. An overall design concept for a system leads to identification of individual components and development of specifications for their performance. This then suggests materials potentially able to meet the performance specifications, either by accessing a computerised data base, by accessing data books or by using in-house experience. From these potential materials unsatisfactory ones are eliminated. Materials passing this screening process (which may be physically applied rather than based on paper studies) now become candidate materials to be evaluated in the design concept. Trade-offs are then made with respect to performance, cost and availability to arrive at a small number of possible materials, and are then differentiated by a testing programme if critical service conditions are envisaged.

In mentioning data books recording numerical properties, it is worth noting that some books are reissued annually and can form part of the reference library for the materials technologist and designer. With metals, for example, one has recourse to the 'Metals selector' issue of the monthly magazine *Materials Engineering*. Once a year, this contains tabulated data only on most of the common, current engineering metals. Published by the American Society of Metals, Volume 1 of the *Metals Handbook* deals only with properties and selection of metals. Its information is incredibly detailed and authoritative. The same Society also produces a two-volume source book on *Materials Selection* which, in addition to presenting detailed numerical information, holds reprints of special papers and original material to meet specific interests.

Metal Progress magazine, also published by the American Society of Metals, produces a regular *Data Book* issue, relevant to materials and to manufacturing technology.

This is just a small selection of support for just one sector of materials. Further information will be found in the Bibliography.

Summarising this sketch of stages in materials selection, the tasks which the product designer is recommended to carry out when approaching a new development are now outlined:

1. Define the storage and service requirements to be met by the product, its materials and associated components.
2. Consider those performance properties important for the product. These may include mechanical properties, chemical resistance, thermal resistance, environmental stability, electrical performance, aesthetics, surface finishing possibilities, produceability, cost of manufacture, maintainability and reliability.
3. Discuss the project with materials experts and product engineers, either in-house in trade organisations, academic institutions, or with materials suppliers and consultants.
4. Search for guidance in the published literature, especially from case histories which involve your proposed material. For plastics, major

suppliers often publish design guidance data books offering specific information on well-defined design problems.

5. Consider again the prevailing attitudes towards your company's past products. Previously conceived ideas on materials performance, costs and availability can quickly become out of date. An unsuccessful brass product of 20 years ago should in no way be allowed to bias your opinion of its possible success in today's market. Seek marketing advice on current customer fashions in material textures and finishes.

6. Undesirable properties of the chosen materials should be identified as they might detract from cost-effective or safe-product performance, or even from overall marketability. For example, in the mid 1980s, any product containing asbestos receives short shrift in the market-place, unless great care is taken to explain that the material is being used in a totally non-hazardous manner.

7. Different classes of materials are subjected to unique testing regimes to obtain numerical performance data. However, these regimes are often not strictly comparable between the classes. Thus, for example, a tensile test made on aluminium alloy is carried out by a method different from that determining the tensile properties of plastics. Quoted single values of a test parameter should not be accepted as the only criterion by which to define a performance envelope.

8. Realistic safety factors and derating rules should be adopted, through a conscious designer decision, to meet customer requirements or to acceed to nationally accepted codes of practice. Safety factors must combine realism with economy. This topic is discussed in a little more detail in Chapter 10.

2.4.2   *Optimum materials choice*

Factors involved in an optimum materials choice have already been mentioned but are not really amenable to a general text such as this. Depending upon the product area concerned, materials chosen may need to be absolutely optimum or may be only approximately so. For example, in the aerospace industry (Table 2.2), a reasonably wide range of very well defined materials is used for different parts of an aircraft, because of the requirements for high performance and absolute reliability, coupled with dependable strength and minimum weight. However, in a domestic product such as a vacuum cleaner, the plastic outer casing may have to meet simultaneous but subjective requirements for strength, rigidity, resistance to solvents, resistance to heat, electrical insulation, warmth of touch, decorative appearance and the rest. No material can meet all the relevant factors optimally, so one reasonable compromise out of many possible has to be made.

When designing with polymers, close attention should be paid to the manufacturing process to follow: for example the injection moulding route to fabricating parts requires special design skills both to understand the process and to achieve optimum manufacturability and performance for the parts.

It has to be recognised that plastic materials undergo time-dependent changes when exposed to steady or variable loads and to temperature cycling. A few degrees change in operating temperature may change the creep rate by several percent. Accordingly, in plastics design, considerable thought should be given to the service temperature envelope to be met, in conjunction with detailed information on static or dynamic loads. After these conditions have been defined, an analysis of the product can be made to define the best manufacturing process, and hence the tooling required, together with positions of mould parting lines, allowable undercuts, the position of injection gates and the general status of inserts, threads and surface finish. Injection moulding as a viable route for manufacturing plastics articles is dependent upon the volume of production foreseen: for relatively small numbers of items, say 100 off, injection moulding would be wildly uneconomic, so other manufacturing processes should be considered.

Finally we should take note of an effect mentioned by Constable (1980) who, after discussing guidelines for selecting materials in products, reported some case histories of materials selection. He found that most industrial companies surveyed could be classified by the materials they preferred to use. They might

Table 2.2. *Some current structural uses of materials in aircraft*

		Aircraft	Helicopters
Composites	Glass fibres	Interior panels	—Blade spars —Rotor hubs —Fairings
	Carbon fibres	Control surfaces	—Blade skin —Rear rotor
	Kevlar	Fairings	Rear rotor
Metals	Aluminium	Main frame	Main frame
	Titanium	Engine blades Engine cowlings	Engine parts
	Steel	Landing gear	Mechanical components
		Mechanical components	

live in the ferrous world, in the non-ferrous metal world or in the plastics world. It is curious to consider why this should be. Can it be that the best materials for a product really are to be found in only one particular group, such as plastics or ceramics? Is it that a past investment in production processes and know-how dictates that the designer must be restricted in his choice mainly to one class of material? Aside from the debate which these questions could engender, the last intriguing point which Constable makes is that most components require a finishing process, but he has found that commercial operations do not necessarily cost out their finishing operations properly, unless of course they are a professional finishing firm providing a service. The designer would do well, if specifying finishes to be applied in-house, to look closely at the economics of this part of his project.

2.4.3 *Value engineering*

This discipline should help the designer choose the best material and design for the lowest final cost. The technique starts by defining accurately the functions which a product has to fulfil and, in so doing, examines the validity of preconceptions about the project. Plastics, being among today's most widely adaptable materials, can often infiltrate areas normally reserved in the mind for more conventional materials when value engineering is properly applied to them. Hall (1981) provides a complex but illuminating text on the potency of the value engineering concept to reduce cost and improve function, and offers as a case history the design of a plastic water meter replacing one fabricated in bronze.

Presumably with value engineering in mind, Reid & Greenberg (1980) describe a programme which used to be run in the Open University in the UK, for design students. During this programme they were requested to design a laminated beam of predetermined length and breadth which, in four-point bending, had to withstand a given force within a specified deflection limit. A wide range of material specimens was provided, together with a table of their relevant properties. The students had to confine their choice of physical dimensions to those of pieces available. The authors discuss the value to the students of this exercise, and also provide equations facilitating design of a beam of a given strength and stiffness; these equations can be solved using a PET microcomputer. The software for this exercise was, at the time of publication, available from the Open University, Milton Keynes, Bedfordshire.

2.5 Data bases and computer aids for materials selection

Selection of material can be effected in many ways other than by reference to the printed page. An American Company has been reported (Anon 1980*a*) as designing its sports equipment and leisure products using a combination of finite element analysis, knowledge of biomechanics and statistical experimental design based upon trials of the product, to decide the optimum selection of materials.

Perhaps more generally understandable are methods depending upon computerised data bases. These may refer only to specific classes of material, as does the CAAS (Computer-Aided Adhesives Selection) technique offered by a commercial adhesives company in the UK. In this simple question-and-answer procedure, the CAAS program guides the user through some 20 situations towards the generic adhesive best suited to his application. The program can be run in two ways: either a full-scale analysis is made to allow progressive examination and re-examination of each parameter to maximise the options or, alternatively, a fast and less rigorous analysis is made which would be sufficient for less critical end uses. Although it is claimed the program can be used by engineers with little previous experience in adhesives, it seems profitable to have the guidance of an expert in the field.

More has been published on data bases for selecting plastics, reflecting of course their wide usage within industry. The Bibliography lists some plastics data bases such as those given by Howard (1980). It seems an obvious exercise to enter pages of materials property information into a computer which is able to respond to inputs specifying required properties and their acceptable ranges of values. Pye (1983*b*) states an enduring fact of design life, which is that for lightly stressed components, thin gauge pressed steel has often been used with much more than adequate resistance to elevated temperatures and service impact. Such superadequacy in service implies a degree of overengineering which should be sufficient to encourage designers to investigate polymeric materials for lightly stressed structural elements. However, because these materials display a less well-defined resistance to temperature and impact than do steels, there is need to choose the most cost-effective plastic. Here we arrive at the starting point for that exercise in computer programming initiated in the UK by Imperial Chemical Industries. Stored basic data on plastics are such that when a set of service requirements is presented, the program will suggest materials to meet them. However, the computer also stores the results of laboratory tests showing how properties vary with time and temperature (creep curves), so that they are combined in what is called a 'design plot', which also indicates the time of onset of prefailure phenomena such as development of crazes and voids. The program can also solve elementary engineering calculations such as deflection rates for loaded beams and flat plates.

More recently, the RAPRA Technology Ltd of Shawbury, England, have announced their PLASCAMS 220 Plastic Materials Selection System. In this microcomputer-based system, the user selects materials on the basis of the various qualities which he would like to see in his product, using menus detailing the properties and production options for plastics in generic groups. It covers not only rigid plain plastics, but those which are fibre-reinforced, filled, lubricated and foamed, and includes thermoplastics, thermosets and thermoplastic elastomers. The end result is often a short list of candidate materials with their typical applications and manufacturers. The search can be carried out serially, that is, materials can be selected to meet first one requirement and then the next.

However, the world of manufactures is not made entirely of plastics, and often more generalised data bases are required. One example, under the trade name MATUS, is offered by the Engineering Information Co. Ltd in London, and is essentially a computerised data base of information on materials and their suppliers, accessible to subscribers via a normal data terminal and a telephone line. Including both proprietary and generic materials, the system allows users to identify materials with specified properties, locate suppliers, and identify materials by trade-name, composition and typical use. It facilitates comparisons across the classical dividing boundaries between metals, plastics and ceramics.

Because of the availability of about 100 000 engineering materials at the present time, of which about 40 000 are metals and 20 000 or so are plastics, the number of computerised data bases held and operated by materials suppliers, universities and technical organisations has begun to increase. Now is the time when rationalised design decisions modelled on the format of these bases should be taken. Bittence (1983) presages the forms of future programs, and the ways in which hardware could be designed to ease searching such data files.

3

An introduction to materials

We have now arrived at the core of the book, the purpose of which is to introduce the designer to the main areas of materials technology. We first consider aspects of the design process which should point up the differences between the performance of classes of materials in practice; and then in later chapters we make a brief review of the main materials properties available. This done, we then consider the 'envelope of properties' of individual materials, which must be defined in order to demonstrate suitability for exploitation in particular designs.

The design process must take into consideration the many aspects of materials behaviour, so that there is never too great a mismatch between the materials actually selected and the realistic performance demanded of the final product. Table 3.1 (Judd 1983) identifies mechanical, thermal, electrical and other performance aspects of materials which have to be considered as a whole or in part during the initial design and materials selection processes. Many of these properties and responses to outside influences are functions of all materials used for engineering designs, although some specifically refer only to metals or only to non-metals.

The properties of materials selected control to a considerable degree, the fabrication methods that can be employed on the ultimate product. Aspects of production, apart from consideration of manufacturing volumes, include the weight of material needed, shapes and sizes of preformed components such as sheet metal and rod, part-size tolerances, and the surface finishes to be adopted. When considering production of metal items, it is usually necessary to consider the machineability of those metals or alloys chosen, even if it is known that they can readily be forged or cast to near final shape. It is also important to ensure that any hot or cold working applied to the metal will impart the necessary extra hardness intended.

Table 3.1. *Material performance factors and their place in materials selection and design*

Factor	Parameters	
Aesthetic	Colour possibilities	
	Optical clarity	
	Surface finish possibilities	
	Freedom to shape in smooth curves	
	Lightness or heaviness	
	Surface properties	
Safety Aspects	Toxicity	
	Flammability and possibility of smoke emission	
	Avoidance of sharp points and edges	
	Shielding of electrical parts	
Environmental	Effect of temperature on properties	
	Degradation due to heat, and to electromagnetic radiation	
	Weathering	
	Moisture permeability	
	Chemical attack from acids, alkalies, solvents	
	Solvent stress cracking	
	Influence of active environments on rupture fatigue, ductile−brittle transitions, and the effect of notches	
	Influence of humidity on creep and electrical properties	
	Biological attack	
	Effect of contact with other materials	
Mechanical	Tensile modulus:	under abnormal dynamic conditions
	Creep and creep rupture:	at various temperatures and stress levels
	Tensile strength	
	Stress relaxation:	at various temperatures and strain levels
	Dynamic stiffness:	at various temperatures, frequencies and stress levels
	Resistance to wear	
	Compressive strength	
	Tear resistance	
	Hardness	
	Impact strength:	notch sensitivity, effect of surface finish, temperature dependence
	Friction properties	
	Dynamic fatigue:	at various temperatures, frequencies and stress levels
Processing	Stiffness	
	Strength	

Table 3.1. (*cont.*)

Factor	Parameters
Thermal	Shrinkage from the mould (plastics)
	Thermal expansion
	Specific heat
	Dimensional stability
	Upper and lower temperature limites in service
	Heat conductivity
	Softening temperature
Acoustic	Vibration absorption
	Damping capacity
Electrical	Electrical strength ac and dc
	Tracking resistance ac and dc
	Volume and surface resistivity
	Dielectric constant at a range of frequencies
	Loss tangent at a range of frequencies
	Conductivity
	Arc resistance
Magnetic	Susceptibility
	Coercivity
	Permanence

When plastics are to be used, the designer has to take care that inserts, under-cuts, holes and stiffening webs are properly catered for in the part design, and consideration must be given to the various production processes applicable to different types of plastic and to the final configurations required. These processes include compression moulding, injection moulding, blow moulding, rotational moulding, extrusion, vacuum forming of sheet, and transfer moulding of thermoset resins, to list but a few.

Many production items are themselves components which subsequently need assembling into a complete product. Some assembly methods have been well established for centuries, such as screwing and hot welding. More recent techniques include various forms of welding, brazing and soldering. Adhesive bonding can be used for joining metals, rubbers and natural materials, and par-ticularly to join plastic materials, which themselves enjoy a large range of other assembly possibilities. These include snap-fitting of flexible lugs on the mouldings, heat sealing, ultrasonic staking, solvent and ultrasonic bonding.

Although much open literature information on materials is confined to specified classes or families, for example metals, polymers or ceramics, other information is regularly published on materials as a whole, to which the interested designer can refer. For example, *Materials Engineering* (1982), presents numerical property information about a wide range of materials used in industrial and consumer products. This includes information on irons and steels, on non-ferrous metals, on plastics, rubber and elastomers, on ceramics, glass, carbon and mica, on fibres, felt, wood, paper, on finishes and coatings, composite materials, joining and sealing adjuncts, and methods of testing and evaluation. In addition, selection charts and comparison data are offered for many of these groups. Essentially, this reference is a Yellow Pages for materials users and suppliers, and despite being devoted entirely to American interests such a document, issued annually, is of considerable use to a designer.

On a more restricted scale, and related particularly to the electrical world, is a shorter document (Anon. 1983*a*) which discusses brazing, soldering and tinning materials, cladding and clad metals, conformal coatings, varnishes, woven tapes, cloth and mat, and heat shrinkable materials.

Accordingly the designer can, if he uses published sources sensibly, obtain expertly condensed knowledge which can increase his grasp of materials technology at a stroke.

Finally, mention must be made in this section of costs as a material parameter. It is, in these days of inflation and industrial flux, difficult to provide a listing of relative costs of materials of construction, an exercise which was far easier in the 1920s. Even if this could be done for longer than the short term, it would not really be too important, because the cost of a product depends on many more factors than that of the raw material going into it. This is true even in the case of precious metals and precious stones, where the craftsmanship of mounting, and costs of cutting, fabricating and decorating can often exceed the intrinsic value of the material, despite the customer's belief that the latter comprises the main investment element.

The total cost of a product or of a component includes that of the material itself (bearing in mind any specialist requirements relating to specified quality, or any discount possibilities through quantity purchased), coupled with the cost of processing the material to the final article. Additional to these costs, also incurred are the costs of inspection at all stages to ensure quality conformance, and subsequent overhead and administrative costs for storing materials and finished goods, for dealing with customers and suppliers and for carrying out subsequent servicing and processing of customer warranty claims. Bearing in mind this formidable list of associated costs subsequent to choosing a material, the designer should be warned that an inexpensive material, which might not be quite suitable for an application, may prove eventually to be very expensive if production time, yield or quality are significantly reduced below expectation had the proper, apparently more expensive, material been chosen in the first instance.

Let us now move on to consider the general properties to be expected of materials for construction, and to discuss some of the constraints on their performance which are likely to be encountered in design and service.

3.1   Properties and performance limitations

This brief chapter discusses the variety of sub-structures within materials which have an effect upon their macroperformance in a product, and then reviews, largely in checklist form, the main attributes that materials exhibit and that can be exploited in designs. It then touches upon two aspects of selection of materials which are receiving increasing attention in modern times: conservation of energy and the ever higher operating temperature requirements.

3.1.1.   *The forms of material for practical applications*

Although, in most of this book, materials are categorised by their main family relationships, so that we speak of ferrous metals, non-ferrous metals, plastics, elastomers, ceramics and composites, materials can also be looked at from the sub-microscopic viewpoint which considers their atomic, molecular and crystalline structures. Engineering materials can broadly be categorised as crystalline solids, in which atoms and molecules are arranged in a more or less regular array as with metals and many plastics or, alternatively, as amorphous materials in which atoms and molecules may have local regularities in their bonding but no long-range overall structure. Such materials often have unique rheological properties and they include some plastics, elastomers and glasses. Obviously, in the engineering world materials that are totally crystalline or totally amorphous are used infrequently. Metals are usually completely polycrystalline unless they are forced to become amorphous by special techniques. Plastics and elastomers may have some amorphous areas adjacent to some of high crystallinity.

The atomic, molecular and microstructural aspects of materials are studied by materials science, which evaluates properties that the designer then exploits in products able to withstand defined service conditions. At this level, a material is considered as an assemblage of discrete particles of molecular size which may be connected together in regular, irregular or random arrays. The materials scientist can interpret such structures to estimate properties likely to be met on the macro scale.

At the structural level, a material can be considered as a composite made up of different coexisting phases. These may be in thermodynamic equilibrium with each other, or they may coexist in an unstable equilibrium liable to change if factors such as temperature are increased; they may, alternatively comprise a mixture of unrelated solids as in fibre-reinforced composites. At this level of material organisation, the materials technologist is able to develop generalised treatments to understand and predict the behaviour of such materials.

While recognising the existence and unique properties of molecular and structural composites, the designer is far more likely to meet with materials at the engineering level which can essentially be considered to be continuous and homogeneous in structure. Average properties may be assumed throughout the whole volume of such materials. This approach dictates evaluation of mechanical properties of large specimens tested under conditions resembling those occurring in an engineering structure during use. The designer can safely accept that the tensile strength of a metal, as shown by a machined test bar, may be reasonably representative of the bulk properties of the material in another configuration. However, a later discussion on plastics and on fibre-reinforced materials will show that, quite often, aspects of the materials structural level must be considered because flow patterns occurring during fabrication can lead to some degree of molecular orientation; this makes the resulting artifact behave rather like a debased fibre-reinforced composite.

3.1.2 General material properties

Some of the exploitable features of materials which a designer has to bear in mind when making a selection from a range of choices, are identified in Figure 3.1. Usually the performance envelope of a material is categorised by a few blanket descriptions of, say, physical, mechanical, thermal, electrical and chemical properties. These are generated by interactions of the structures present at molecular and structural levels. The designer has to select materials whose properties can be optimised, given the range of structures which the material can be made to adopt from its initial manufacture or through modification by processing. Thus, for example, the yield strength and ductility of a metal appear very different if it is tested in the fully annealed form, as it might be when cooled from a melt, compared with the strained form which will occur after it has been subjected to a bending process for final shaping. Thus the designer must be certain that he understands, at least to a limited extent, the effects of processing conditions on the structures which are present or which can be generated in materials, since quite often 'to bend' is 'to change physically', and this is reflected in a change of several properties. We are all familiar with the phenomenon of work-hardening of metals, and of the fractures which occur when a piece of sheet metal is bent repeatedly through a wide angle. Bending the material once or twice may, however, confer a useful degree of strength to a product; the possibility of this happening must be recognised and capitalised upon.

The principal methods for altering structure in metals are adding alloying elements to change the composition, applying heat treatment to dissolve or preciptitate phases in the structure, and deforming by processing in the way already mentioned.

A caution should be sounded here on the quality of experimental data. Reliable and reproducible test data on materials are requisites for correlation of theoretical and experimental evidence for establishing the validity of a design. However, the quantitative value recorded for a property may vary depending upon a particular test method using specific equipment; specialist expertise may be required to conduct the test and to interpret the results. For example, if one takes two geometrically similar bars of metal and plastic, and tests for tensile strength by standard methods considered appropriate, then the regime for testing the metal sample will operate at a lower strain rate than the test for the plastic one. If both were tested at the same low strain rate, the metal would ultimately fail under tension, whereas the plastic would relax and may not even fail at all within the geometric and time limits of the test. Because testing regimes are different, numerical values obtained for a parameter are not strictly comparable; the designer should be aware of this fact even if he cannot directly use it in his design.

Identification of the most appropriate physical property data to meet specified requirements is quite important. Many standard testing methods for

Fig. 3.1. Main exploitable features of materials. (Copyright: 1985 Standard Telephones and Cables plc.)

	Mechanical				Thermal			Electrical		
	Strength	Hardness	Density	Elastic mod.	Melting point	Expansion	Conductivity	Conductivity	Dielectric strength	Dielectric constant
Metals	✓	✓	✓	✓			✓	✓		
Polymers	✓		✓			✓			✓	✓
Ceramics	✓	✓			✓		✓		✓	✓
Composites	✓		✓	✓						✓

materials were originally developed for quality control purposes; consequently test conditions were devised for ease of operation and repeatability rather than for meeting the requirement for which the results are now needed. For example, an elastic limit represents the first significant deviation from true elastic behaviour of a metal. However, measurement of this parameter is difficult and specialised; so instead of determining the real elastic limit, one usually measures and subsequently quotes the 0.2% offset yield strength value. This can be used for engineering design purposes but on the other hand it, too, because of the need for a machined test specimen, may be inconvenient to use, so an approximation to the yield stress is finally made by a rapid and inexpensive hardness test. The designer must be aware of this chain of events so that he can, in some way, mentally evaluate and judge the disparity between choosing his material on the basis of a hardness value, when really he is interested in its performance under cyclic tension conditions, and presumably would like to know the true elastic limit.

Returning to the theme that test results, as explained in property data tables, may not represent the full truth, it must be realised that in the stable, settled and orderly life of the laboratory it is desirable that tests should be made under reproducible conditions. Thus a tensile or ductility test may be made under steadily increasing strain at some conveniently chosen linear rate. However, in real life, tensile forces on a structure such as a steel railway bridge may well be erratic, spread over a range of values, and randomly cyclic to boot, in addition to operating over a temperature range of maybe 50 °C. Such forces, if applied for long enough, could lead to fatigue failure or creep, which really puts the relevant testing regime into a totally different area from that which the designer may have visualised.

So the message of this homily is that the designer should think carefully of the use and service conditions of the product with which he is concerned, and should consider adopting results from tests which most nearly approximate those conditions. Of course this will not always be practicable if only for expense, availability and time reasons, but the designer may very well, through his own experience, eventually find those parameters best suited to his needs.

3.1.3. *Energy considerations*

Energy and materials in world economics are inextricably tied together. Basic materials are won from the earth by expenditure of energy, and purified by using more energy, before being transported and fabricated into final products, when yet further energy is used. A survey by Mesker (1978) implies that about 10% of the total cost of making an engineered part was then attributable to the cost of energy locked in it; the people surveyed asserted that if energy costs rose to 12%, they would change manufacturing processes and make other moves in order to conserve it. The general tenor of the Mesker survey was that proper control of manufacturing processes can reduce costs through energy savings.

Resource conservation has been a major topic of discussion since the 1974 oil crisis. Seen from a national viewpoint the relationship between materials production and energy consumed during their manufacture is significant for an economic balance; it can affect strategies for winning or stockpiling materials. The specific energy, and energy per unit of tensile, modulus and fatigue strengths for some common metal alloys, plastics, concrete, glass and timber, have been calculated and analysed by Alexander (1979) and the cost related to materials properties. His analysis of the range of materials concludes that timber, the cheapest from an energy consideration, becomes almost the most expensive when providing a specified mechanical performance. A high cost per unit property is also apparent with stainless steel and with titanium, presumably because of the high costs of smelting, casting, working, surface finishing and maintenance of purity. In the short term, energy savings for the world are made possible by improving processing efficiency, by concentrating on materials which can be recycled from scrap without deterioration of properties, or by using those materials which have a low total energy content such as timber, concrete and most steels. For any enterprise concerned with cost reduction, the total energy content per unit of property, combined with the material cost per unit of property, should be used as one basis for selecting materials. More specifically, in a production unit concerned with injection moulding of plastics, the steps needed to reduce energy usage in plastics injection moulding operations are well defined. Anon. (1981*a*) supplies a checklist of items which should be evaluated, and offers some algorithms for improving machine settings and controls.

3.1.4. *The problem of operating temperature*

One of the most critical questions in materials selection is the operating temperature of the product during its service life. A recent design package Knight (1983*c*) relates the cost data for many engineering plastics to their maximum service temperature. Using these data, designers can identify the additional cost if demanding, for example, an extra 10% of temperature resistance. Of course, selecting a material on the basis of service temperature alone is not going to be any more safe or constructive for the integrity of a product than if any other single parameter is chosen. If a candidate polymer displaces another in a designer's mind because it operates at a temperature 10 °C higher, other constraints will inevitably show up; for example, the manufacturing processes may prove more difficult and accordingly more expensive. Knight explains in tabular form how the volume cost per operating degree Celsius above room temperature varies between engineering thermoplastics, which can be ranked in increasing order.

Saillant, essential

3.2 Salient features of main classes of materials

In a broad and very general way, the main materials available to the designer for engineering applications can be divided into four classes comprising metals, polymers, ceramics and composites. Materials commercially and technically known as polymers are mostly carbon-based chemicals with long molecular chains. Ceramics, on the other hand, are based on chemical elements of other kinds and may in some cases be derived on a molecular scale from very large assemblages of atoms. Composites also possess the properties of any of the other three classes insofar as the matrix material may be a metal, polymer or ceramic, and the reinforcement may be particles, fibres or cloth.

Each of the main classes of material has its own well-defined 'envelope of properties', the broad features of which are shown in Table 3.2. An alternative way to view the main materials classes is to assess those key properties which tend to be exploited time and again in designs. Some of these are illustrated in Figure 3.2 and suggest that whereas in metals, the strength, hardness and modulus can often be taken for granted, sometimes their high (or low) density might be an advantage in their use. Having an electronic structure, metals conduct heat and electricity quite well. On the other hand, polymers, although they may be sufficiently strong, can often be exploited simply because their density tends to be low, as is their heat and electrical conduction capability.

Although not always an advantage, plastics thermal expansion has to be taken into account because of their high values, compared with those of more traditional materials. Polymers are also valuable to the designer on account of their dielectric properties. Ceramics are usually strong and hard and are often

Table 3.2. *Characteristic envelopes of properties for the main materials classes*

Class of material	Salient properties
Metals	Stiffness; strength; ductility; hot/cold working; machineability; weldability
Polymers	Easy processing; mechanical properties sensitive to temperature; deflection sensitive to temperature, time and strain rate; effect of environment and solvents
Ceramics	Strong in compression; weak in tension; brittleness; shrinkage on firing is significant
Fibre composites	Anisotropic properties unless fibres are well oriented; high strain in fibres can result in large deflections; low fracture toughness; care needed in joining components

exploited for their high melting point and reasonable thermal conductivity values. Composites, on the other hand, are often specified because of their combination of high strength and elastic modulus, coupled with low density.

A considerable number of material properties can crudely be grouped into mechanical, thermal, electrical, and other attributes, and a range of these is shown in Figure 3.2. By deploying an adequate general knowledge of materials technology when considering such a diverse list of product requirements, the

Fig. 3.2. Specific requirements which can be met by materials. (Copyright: 1985 Standard Telephones and Cables plc.)

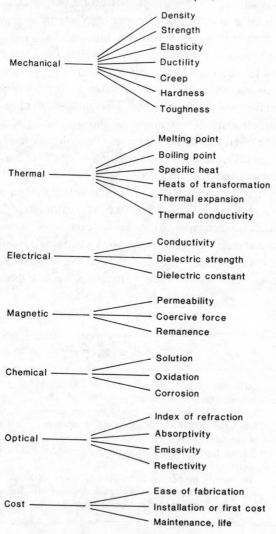

designer has some hope of finding a suitable base material, variations of which might then be adapted to the specified product requirements, or at least to most of them.

We should now look at some of the salient features of the main classes of materials. Metals in the usual polycrystalline state, apart from displaying in good measure the attributes of Table 3.2, are not easily amenable to alteration of their basic elastic moduli by heat treatment or by other processes. So if a component for a product must have high stiffness, this can only be made from metals by choosing a stiff alloy, or by changing the shape or profile of the part in some way. The range of strength and stiffness values achievable in at least some alloys, compared with some fibre composites, is demonstrated in Table 3.3.

However, heat treatment of many metals can change strength and ductility either separately or together. For example, annealing will usually increase ductility by dissolving second-phase metal grains. On the other hand, heating may cause recrystallisation and hence precipitation of a secondary hard phase as in some aluminium alloys, and so increase the strength. Metals also cause problems for the production engineer insofar as heat treatment or cold working

Table 3.3. *Some mechanical characteristics of 'high strength' metals and composites*

Characteristic Material	(1) Static strength (MPa)	(2) Specific mass (kg/m^3)	[1]/[2] Specific strength (m^2/s^2)	(3) Young's modulus (MPa)	[3]/[2] Specific modulus (m^2/s^2)
Light alloy 2024	420	2770	150 000	77 000	27 000
Conventional steel	1100	7800	140 000	210 000	27 000
High strength steel	1800	7800	230 000	210 000	27 000
Unidirectional carbon fibres + epoxy resin	1400	1560	900 000	130 000	84 000
Isotropic carbon fibres + epoxy	400	1560	260 000	50 000	32 000
Unidirectional glass fibres + epoxy	1400	1900	750 000	42 000	22 000
Unidirectional Kevlar + epoxy	1450	1370	1 050 000	87 000	63 000

After J. Balazard
SNIAS France, 1984.

Table 3.4. Comparison of some structural material properties

Property	Units	Al		Strip steel	Stainless steel	Steel grade 5 43/50	Glass Reinforced polyester	Plastics/ elastomers	Timber
		Cast	Wrought						
Specific gravity		2.65–2.68	2.69–2.73	7.84	7.7–8.0	7.85	1.4–2.0	0.9–2.0 (unfilled)	0.5–1.1
Thermal conductivity	W/m K	138–151	109–201		15.25	50	0.17–0.51	0.17–0.69	0.13–0.15
Thermal expansion coefficient	$\times 10^{-6}/x$	20–23	23.0–24.5	10.0 –12.6	10.5–18.0	12	4–36	54–336	4–6 parallel to grain; 30–70 across grain
Maximum temperature	°C	100: seek advice if higher			750–1100	550–1000 (Depends on load and exposure)	260	90–260	100
Specific heat	J/kg K	920		450–481	490–520	450	1250	1260–2520	1210
Minimum service temperature	°C	–200	–200		–273	Depends on grade		–50	
Modulus of elasticity	$N/mm^2 \times 10^3$	71	69–71	205	195–205	210,000	5–10	1.4–15	6–17
Specific stiffness	$N/mm^2 \times 10^3$	27	25–26				3.6	1.5–5.9	14–31
Yield strength	N/mm^2	90–240	70–270			245–355	Tension 80. Compression 20.	6.9–27.6	
Tensile strength	N/mm^2	140–310	112–375			430–490	70–1150	20.7–103.0	3.4–41.0

Elongation %	1–13	4–25	20–22	2.5	100–400
Hardness Brinell	55–105	29–110	120–150	45 (Barcol GY21 934–1 test condition)	M40–M120
Rockwell M scale				110	
Impact strength (notched Izod) J/cm²	3.7	4.8	27(Charpy depends on grade and temperature)	7.5–25.0 (unnotched)	0.26–3.7
Compressive strength N/mm²	280–590		430–490	200–480	694–60

can lead to stress-induced dimensional changes, which may result in tolerance problems later. Machineability of metals varies widely, as will be seen in Section 4.4. Weldability of alloys is affected by their composition, an example being some steels which, if they contain too much carbon, can suffer undesirable transformations of the metal structure which will lead in turn to reduced strength of the finished product.

The key consideration when choosing polymers for a product is that mechanical properties are sensitive to time, environment, temperature and applied strain rate. Numerical values of some mechanical properties can often be increased by incorporating fillers such as glass fibres or particulate powders, or reduced by incorporating air to generate a foam. Generally the envelope of properties of polymers lies in a range of temperatures much lower than that of metals. Polymers share with other organic materials chemical changes caused by exposure to ultra-violet light, oxygen, fire, high temperatures, water or solvents. However, from a production point of view, they are fairly easy to process by a very wide variety of methods. Because of the inherent simplicity of the injection moulding process, very complex shapes can be imposed on a single part. This can often be exploited to eliminate the assembling of a number of sub-components which might be needed if the part were made of metal.

The key feature of most ceramics is that they are particularly strong in compression but usually quite weak in tension, because of the presence of microcracks. These cracks act as stress raisers in tension and thus soon lead to brittle fracture at low strains. Cracks may be encouraged by poor design, for example by allowing too small a radius at a section change, or may occur naturally in a product through accident or bad production processes. Ceramics, which are always finally fired to achieve densification of the originally powdered material, will mostly suffer significant shrinkage during this final stage. Shrinkage may not always be isotropic and for engineering purposes has to be compensated for in the design. If fine tolerances are needed, the fired product must be manufactured oversize and then shaped afterwards by wet grinding, air abrasion or by some other machining process.

Composites, particularly those which are fibre reinforced, achieve their high strength largely because stress applied to the matrix material is transferred to the strong fibre reinforcement, provided that a good chemical and physical bond occurs at the interface between the two. Although high strains can be transferred in this way to give structurally sound and mechanically resistant products, acceptance of strain might be at the expense of an unacceptable deflection. As an example, it has been calculated that an aircraft wing made entirely of fibre-reinforced plastic might, under a 1.7% strain, deflect so much that its tip becomes vertical! However, it should be noted that aircraft wing designers normally resist specification above a maximum of 0.9% strain. With fibre composites, the not-so-random orientation of fibres, unless carefully controlled by the manufacturing process, leads to anisotropy of mechanical properties, although in some applications such as epoxy-glass flat leaf springs, this feature is positively exploited. As a generality, fibre composites often exhibit reasonably high fracture toughness, and compression loading tends to failure by buckling and delamination, although both these problems can be designed out once recognised. Joining of composites by mechanical means can offer problems because localised stress concentrations can be set up, leading to separation of the matrix and the fibre at the fastening point.

It might at this stage be as well to study Table 3.4, which provides comparative figures relevant to some common classes of structural materials. Ranges of values are offered because obviously different types of materials from unspecified sources will exhibit some variation. In a few cases the variation appears excessive (for example, the yield strength of wrought aluminium varies from 70 to 270 Newtons per square millimetre). However, these merely reflect the variety of products which the designer is able to exploit.

Before proceeding through Chapters 4,5,6 and 7, to review individually the main classes of materials, it is probably helpful to consider in a little more detail how to evaluate the materials envelopes of properties briefly mentioned in Table 3.2. The general problem lies in relating known performance requirements of the product with tabulated or graphical materials data, in order to achieve a reasonable match.

One method by which this can be done is shown in Figure 3.3, which lists in columns the most common relevant properties. For example, (1) might relate to numerical tensile strength values, (2) to numerical impact values, (3) to colour capability, (4) to flammability potential, and so on, all quantified on some chosen basis. These columns are chosen in number and nature to match the performance requirements expected of the product under consideration. In the rows, potentially suitable materials A,B,C,..,X,Y,Z are listed; the matrix is completed by inserting at least approximate numerical values for each material in the property column in question. Materials listed in the rows may be disparate; perhaps 'A' may be copper, 'B' may be polyamide plastic, 'C' may be oak, 'D' may be concrete, and so on. Often the choice will obviously be much more close run than in this example and the materials in the rows may be closely related, for example being varieties of aluminium alloys.

Having decided the percentage deviation from product requirements permissible in the materials rows to allow an acceptable selection, the figures in the body of the matrix are ringed if they fit the performance requirements reasonably well. It should then be possible by looking at the number of acceptable properties along any row to decide which material most nearly meets the performance requirements.

Proceeding beyond this approach, we might assume that two different materials each seem a reasonable choice for fulfilling the product specification. We can now draw graphical property profiles, as shown in Figure 3.4. In these the required key properties are drawn as lines on the ordinate, while the abscissa indicates the percentage range of each property which can be exploited in the product while still retaining some semblance of its original performance expectation. By connecting the ends of the property range lines, a literal 'envelope of property limits' is obtained as in the lower two diagrams for materials 'A' and 'B'. Similarly, as shown in the top diagram, the envelope of properties optimally required for the product is illustrated. The middle two diagrams show in what respects materials 'A' and 'B' match the product requirements. This can easily be tested in a practical case by drawing each materials envelope and the required product property profile on separate transparent foils, superimposing them to look for areas of commonality. In this particular example, the reader must himself decide whether material 'A' or

Fig. 3.3. Tabular method for matching materials properties and product requirements. (Copyright: 1985 Standard Telephones and Cables plc.)

Property reference

	1	2	3	4	--------	42	43	44	45
Product values (± x% deviation)	17	21	40	109	--------------	84	No	17	120
Possible materials A	(21)	60	103	84	----------------	68	(No)	24	110
B	11	(19)	84	136	----------------	(82)	(No)	29	96
C	407	120	13	8	--------------	70	Yes	13	108
D	(14)	41	(45)	135	--------------	91	Yes	(15)	(124)
E	(18)	(20)	(32)	16	--------------	(82)	(No)	21	113
F	61	110	8	34	--------------	68	Yes	9	131
G	76	28	52	(111)	-----------	90	(No)	(17)	(120)
X	(18)	42	60	126	-------------				
Y	26	32	50	94	------------				
Z	9	(23)	(34)	(115)	-----------				

Fig. 3.4. Matching product envelopes of properties with materials envelopes. (Copyright: 1985 Standard Telephones and Cables plc.)

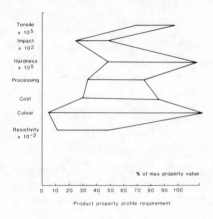

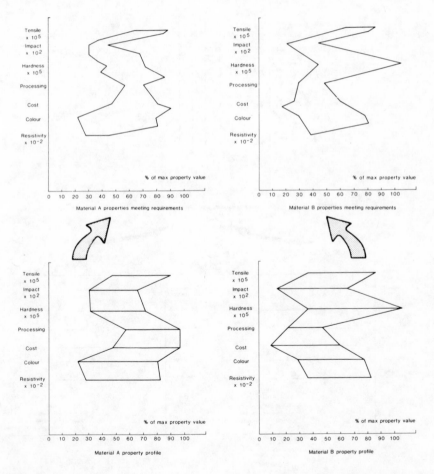

Fig. 3.5. Alternative materials property envelope presentation. (Copyright: 1985 Standard Telephones and Cables plc.)

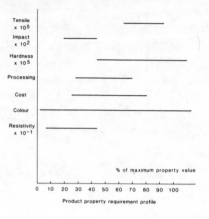

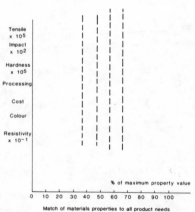

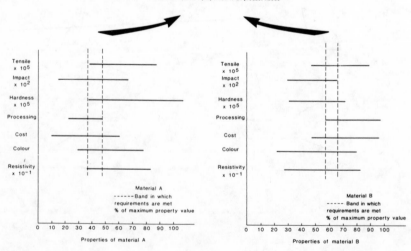

material 'B', as shown in the middle drawings, most nearly matches the requirements of the product property profile.

An alternative presentation method shown in Figure 3.5 lists the properties of concern on the ordinate, with a 0-100 scale on the abscissa. This scale is not absolute in value but defines the limits for any one of the parameters; thus when considering the tensile properties, the scale might range from 0-1000 N/mm^2 and the same scale for the impact parameter may represent 0-10 J/cm^2. Processing would be defined as a specific number outlining each process, so would essentially constitute a series of spot points or bands.

Fig. 3.6. Matching seal performance requirements with elastomers.
(Copyright: 1984 EI DuPont de Nemours)

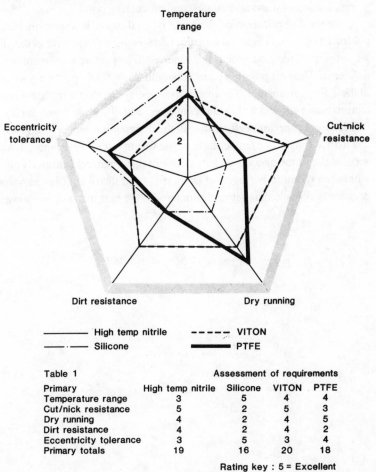

Table 1		Assessment of requirements		
Primary	High temp nitrile	Silicone	VITON	PTFE
Temperature range	3	5	4	4
Cut/nick resistance	5	2	5	3
Dry running	4	2	4	5
Dirt resistance	4	2	4	2
Eccentricity tolerance	3	5	3	4
Primary totals	19	16	20	18

Rating key : 5 = Excellent
1 = Poor

In this diagrammatic representation the interest is in the extent to which the properties of materials 'A' and 'B' simultaneously meet the product property profile shown at the top of the diagram. Vertical parallel lines define a band over which at least some aspect of every one of the required properties is reflected by the candidate material's performance. Comparison of these bands in the centre diagram shows that material 'A' broadly meets the requirements of the product profile, since most of the parameters are met by it. It does not quite meet the tensile strength demanded, but is much better on impact performance than was specified. It meets the other parameters simultaneously but has a rather high resistivity value. These considerations, and the narrow band of congruent properties, make material 'A' a real contender. On the other hand, material 'B' shows a broader match of all the properties, but an overall higher performance. In such a case, a decision between the two might well be based on a combination of material first cost and processing costs.

A practical application of this type of evaluation is shown in Figure 3.6 published by permission of the DuPont de Nemours Company. In this diagram we see a pentagon on which the required five performance parameters are indicated. The best possible material would show 100% performance for each of these five parameters. The materials property envelopes for four contending materials are drawn within the pentagon, and illustrate how, for example, PTFE offers superior performance in the dry running aspect, but is less effective than Viton or high temperature nitrile for cut or nick resistance. Taking into account the virtue of concentricity of one of these materials property envelopes within the required pentagon of properties, and the overall larger values, the decision was made that Viton elastomer was the best material to choose.

4

Properties of metals and alloys

Metals pervade industrial life and are the obvious first choice for making industrial, professional or consumer durables of small to enormous dimensions. Products of metal range in size from a pin to a major suspension bridge or an ocean-going liner. Having said this, it is useful to understand the realities of sources of metals and their general supply characteristics, both in terms of world raw material availability and in terms of stock materials obtainable from a metals supplier. Varjian & Hall (1984) review the modern situation in some detail. A combination of high metal stocks in the London Metal Exchange warehouses, and a slow economic growth worldwide has conspired to put down the pressure on prices and hence to depress the staffing levels and profitability of mining and mining-related jobs in many countries. The authors then consider the mid-1980s situation for several metals.

In 1983, world production of aluminium was about 14 million tons of which half originated in the Americas, prices being around $0.8 per pound. Chromium is largely produced as chemical compounds, the amount of chromium-based alloys and basis metal being relatively small in the light of an annual South African chromium ore production of 2.4 million metric tons. The copper industry in 1983 continued to suffer from world oversupply of the metal, of which almost eight million tons were produced in the year, at prices of the order of $0.75 per pound. Magnesium enjoyed a resurgence of interest in 1983 with a world production of over a quarter of a million tons, reflecting increasing application awareness by the automotive industry. In contrast to the foregoing tonnage amounts, nickel production in the world is reckoned in millions of pounds per annum, just over one thousand million pounds being produced in 1983 at prices of the order of $2.25 per pound. World production of zinc totalled over six million tons in 1983 at prices of almost $0.5 per pound at their peak.

Enough has been said to show the scale of manufacture of some of the common metals specified by designers, ignoring totally the enormous world steel production assumed as adequate in terms of base metal supply and the variety of alloys which are offered. Metals and alloys are normally bought from suppliers as castings, or as wrought material in preformed stock shapes. Preformed sections are available as bars, rods, tubes, strips, sheets, plates and coil stock, and, in the case of some softer metals, such as aluminium, in the form of extrusions with cross-sections which can be specified to suit a particular design. Preformed materials often offer the easiest, and hence the cheapest, route to a desired product shape; designers should exploit their possibilities.

When considering metals for product applications, costs per unit weight and volume are frequently compared. The latter is the more significant comparison because of wide differences in density between metals. Furthermore, consideration of volume brings one to the cross-section of, for example, beams, and this enables calculation of specific strength per unit of cost, thus allowing comparisons between what might otherwise be widely different contenders for a specific application. It is pointless to offer here cost comparisons of the range of metals likely to be of interest to the designer because economic factors come into play too frequently. For example, under conditions where there happens to be a world shortage of molybdenum, nickel − chromium − molybdenum stainless steels, such as Types 316 and 317, are sold at a much higher premium than the nickel/chromium stainless steels which at other times could be of comparable price.

Another important factor is the cost of material fabrication. In making hollow vessels, for example, cutting and welding of plates may involve labour and handling charges in excess of the cost of the basic metallic sheet. In a typical carbon steel pressure vessel, the fabrication may cost more than 50% of the total. Similarly in piping systems, formation of bends and welding of flanges may dominate the costing. Generally it can be asserted that because of fabrication costs, small differences in price between materials of similar grade will not significantly affect the final cost of the product. Differences in fabrication costs may be even more marked when one is deciding whether to use metals or, for example, plastics for a given construction. For the less expensive metals such as carbon steel and aluminium, fabrication charges represent a greater proportion of the total cost of an equipment than the basic material. With more expensive alloys such as those with a nickel base, material costs can predominate, but even so their effects may be smaller than expected.

There still exists a world problem of profligate waste of metals during processing and use. In any society other than one which completely recycles all the mined metals (and this excludes every society in history including our own), the

net result is that metals are mined, used and then distributed by man in a thin ir-
recoverable layer over the surface of the occupied world. Aluminium drinks
cans spread alongside roads are a tangible symbol of this situation. This means
not only that mined resources ultimately become totally and irrecoverably
depleted, but the thin spread of material over a large surface area precludes any
practicable, let alone economic, system of recovery. A consequence of this
unrelenting process is that contamination of crop-growing areas and of people
continues from the small but important amounts of toxic metals being spread
everywhere. A significant proportion of the value of these metals so wasted lies
in any non-renewable energy invested to win them in the first place. Some small
measure of economy can be achieved by using less energy while wasting this
metal. Chadwick (1984) asserts that best economies can be achieved by near
net-shaped forming. This means that if a metal could be compacted and produced in the final shape of a product without the intervention of machining
processes, then swarf, with its own investment of energy entirely wasted,
would not be produced in the first place. Near net-shape forming processes
include powder metallurgy, which is likely to have a minor impact as artifacts
produced are inevitably geometrically small, and rheocasting which implies
die-casting under pressure a partly melted thixotropic metal alloy amenable to
total scrap recovery and reuse. Another process applicable to fibre composites
is slush casting; lastly there exists a whole range of die-casting processes
intended to reduce metal use and wastage by producing castings free of pores
(except for aluminium, which makes inherently porous castings) and hence of
structural weaknesses. Such processes include permanent mould casting,
investment casting and sand casting. Another possibility, as yet unexploited on
any large scale, is that of making metal shapes by spraying from melted metal
with subsequent pressing to final form. Chadwick also points out that there is a
high correlation between the world use of energy and the extraction of
materials. At present, $50-60\%$ of the world's energy used is involved in extracting metals from the earth and converting them into usable form.

Metals as a family are opaque and lustrous, comparatively denser than most
other materials and often strong, malleable and ductile. They tend to conduct
heat and electricity well in comparison with other materials. Mechanical
properties in particular relate to the metallic characteristic of packed crystal
structures favoured on the atomic scale by strong metallic bonds.

It is easy to speak glibly of 'advantages' or 'limitations' of a material, but
materials have no such inherent attributes; these are entirely man-made.
'Advantages' and 'limitations' are reflections only of the designer's expectation of what he would like to exploit in a design, or of what he will have to live
with as a consequence of selecting one metal in place of another. For this

reason, and because of the limitations of the size of this book, brief comments will be made on only a few metals, to sketch in a few characteristics which should be recognised if the designer is to develop an intuitive feel for them.

Steels tend to be relatively very ductile, heavy, non-combustible with a high-strength-to-weight ratio and of reasonable stiffness. They are amenable to formation into a wide variety of shapes by bending, drawing or forging. The most obvious problem with common carbon steels is their tendency to react with moisture and oxygen to form oxides more bulky than the metal itself. These oxides or rusts swell up, drop off and fresh material again reforms on the newly exposed surface, which results ultimately in the steel article being totally consumed unless protective systems are applied to the surface or the ambient atmosphere is properly controlled. The presence of moisture and other metals can also promote preferential local corrosion, usually to be overcome by joint design or by appropriate cladding of the metals concerned. Stainless steels overcome many of the problems inherent in the performance of carbon steels, especially in respect of exposure to corrosive environments and to heat. However, there are many varieties of stainless steel, each with its own envelope of properties. Designers are urged to consult a competent metallurgist for detailed advice before making a firm choice.

Another major difficulty with steel work is its reaction to fire. Although the metal has a melting point well in excess of 1400 °C, in a fire situation steel structural members suffer a severe loss of strength and effectively are totally annealed at temperatures as low as 600 °C. The result is to be seen in any steel-framed building which has suffered a serious fire without adequate protection of the structural members; the metal deflects strongly under modest applied loads.

Aluminium, on the other hand, in its unalloyed form, is highly resistant to atmospheric corrosion. However, the pure material is very soft and so is not used for structural purposes. The more highly the aluminium is alloyed with strengthening elements, the more likely in general it is to corrode; for this reason alloys need careful selection, preferably by a metallurgist, for the proposed use conditions. Low density means that lightweight constructions are possible. Although aluminium retains useful mechanical properties at very low temperatures, the properties as measured at room temperature are reduced by exposure to temperatures above 100 °C which is often a low performance requirement. Relative to steel, aluminium is costly and thus very large members are less common than those produced from ferrous metal, despite the advantage of lower density. Other advantages of aluminium include its easy working into thick or thin sections by extrusion, casting, profiling, and by many other means, and the possibility of improving resistance to wear and corrosion by anodising, although alloys for such treatment need careful selection. The material is easily welded. Disadvantages compared to steel include a relatively low modulus and a rather low fatigue endurance.

Magnesium and its alloys are the lightest of all commercially available structural metals, and compare favourably to aluminium on a strength-to-weight basis. Alloys are available as castings but care is necessary when choosing them, as ductility is limited in comparison to aluminium. Hot working is possible, usually by extrusion, and the alloys do not attack the extrusion machine tooling parts as much as aluminium does. Magnesium components suffer a limitation through poor resistance to the environment, and a surface coating is usually needed to maintain structural integrity. The upper service temperature is of the order of 300 °C.

Zinc is seldom used alone as a structural material, but the pure metal is applied as a coating to steel by electroplating, dipping or spraying. Zinc alloys are used extensively as die-castings and offer good reproduction of complex shapes in the die, coupled with low casting temperatures which lead to attractive economics. Castings have a good corrosion resistance in natural atmospheres, but for decorative purposes are capable of accepting a wide range of surface coatings. It is claimed that zinc die castings are stronger in many ways than plastic mouldings, although this differential might be eroded by further developments by the plastics industry. Main limitations of zinc die-castings are a low specific stiffness and modest resistance to creep. Corrosion occurs in any acid or alkaline environment which departs very far from neutral conditions.

A comparison of some metals properties provided in Table 4.1 should enable the reader to relate his own experience with what might be expected from metals that are unfamiliar to him. The table exhibits approximate values only of significant properties; because of the wide range of alloy compositions available it is not possible to be more definite or accurate. Full data on specific candidates for a design should be sought from the materials supplier.

4.1   Aspects of selecting metals and alloys

The world is full of metal artifacts of one kind or another, ranging in size from (almost) microscropic screws as used in spectacles, through vehicles of all sizes to bridges, ships and steel framed buildings. Although defects occur in service, being usually due to human errors in design or commissioning, nevertheless, the high success rate for the use of metals reflects their secure place in the designer's armoury, since a tradition of experience over thousands of years has been built up, much of which seems to be in-built into engineer's subconscious. So space will not be wasted here in discussing successful applications where one metal was chosen instead of another. Brief reference will, instead, be made to those aspects of selecting metals which must be borne in mind if the ultimate product is to have some chance of success. Good design and

Table 4.1. *Comparison of some metals properties*

Metal or alloy	Specific gravity	Melting range (°C)	Yield point (N/mm²)	Modulus of elasticity (kN/mm²)	Ultimate tensile strength (N/mm²)	Elongation % on $\sqrt{(5.65)}$ s_o	Hardness (Brinell unless marked)	Izod impact strength (J/cm²)	Electrical resistivity (μ ohm-cm)	Thermal conductivity (W/m K)	Coefficient of linear expansion (×10⁻⁶ K)
Aluminium (castings)	2.57–2.81	477–649	90–240	69	240–310	2–8	65–90		8	138–151	21
Aluminium (wrought)	2.66–2.84	449–649	70–270	69	112–375	6–40	20–70		4	109–201	23
Brass Cu–Zn (60–40)	8.47	932	108	103	151–324	4–55	65–185 (Vickers)		7	125	19
Copper	8.9	1082	108–324	117	96–172	4–60	45–115 (Vickers)		1.7	408	17
Iron, ductile cast	7.2	1149		172					60	34	13
Lead	11.35	327		14					21	36	30
Magnesium	1.8	650	77–154	41	69–255	1–15	35–70 VPN 50–80 (Brinell)	0.26–1.37 (Charpy)	4	89	29
Nickel	8.89	1441		207	180	15			10	63	12
Phosphor bronze Cu–Sn–P (63–37)	8.98						(110 Rockwell)			91	32
Solder Pb–Sn	8.89	183							15	48	24
Steel (0.4% carbon)	7.85	1515		207	228	28	135	3.42	10	55	11
Steel (cast carbon)			185–587	207	193–310	8–22	170–300	5.26–10.5	17		11
Steel (mild 0.6% carbon)	7.87			207	207	45	92	24.2	18	70	13
Steel (stainless)	8.02	1427		193					72	19	17
Tin	5.77	232			6.9				12	36	36
Zinc	7.14	418							6	113	32

correct materials selection can greatly reduce the incidence of corrosion or fatigue failures in subsequent service. Consultation between designers and metallurgists at the drawing board stage has the potential for making untold savings. ~~Eminont, dominant~~

The pre-eminent property which metals offer over other new or traditional materials, is that of consistently reliable high strength. While this of course is relative, insofar as lead, for example, is a weaker material than some timbers, nevertheless in general for constructional purposes one looks to metals to offer high tensile strength and other mechanical properties in good measure. Usually the tensile strength of a metal is an academic parameter, since if its value were reached in service, the product would be irremediably deformed by elongation, although design guidance has been reported to encourage this effect to equalise stresses in a complex structure. More practicable for design purposes is consideration of the yield strength or the $0.1-0.2\%$ proof stress, below which the metal acts in a truly Hookean (elastic) mode. Like most statements, this is not entirely true since some metals show hysteresis effects, but these are likely to be small for the majority of 'normal metals' with which designers must deal.

Another aspect of metals is the general property of toughness, which represents the resistance of materials to brittle fracture (mostly brought about by previous fatigue exposure) on impact. Here, too, metals run the whole gamut of performance from being rather brittle, as in the case of cast irons, to extremely tough, as in the case of some manganese-based ferrous alloys used for plough shares and dredger teeth. However, impact resistance is dependent upon other factors, such as work hardening and operating temperature. Work hardening can be developed in some metals, either as a byproduct of normal manufacturing processes, or specifically applied, as in the case of rolled strip steels, drawn rod and wire, as a means of upgrading material performance. Work hardening offers a strength increase in terms of tensile and yield parameters, but always reduces ductility. Strength increases and ductility reductions also follow a decrease in operating temperature, so that selection of metals for cryogenic applications becomes a very critical matter indeed.

Perhaps the second most important aspect of metals selection, which the designer must address, is that of cost. The cost of untreated mineral ore may be quite low, as in the case of iron, but additional costs are incurred in winning the metal from the ore, purifying it and applying refining and shaping processes, until it reaches a form suitable for the market-place. Having achieved this, it is necessary to consider the cost in relation to the strength offered. A conventional design parameter allows for the effect of density by comparing costs of metals on a unit volume basis, or on the basis of a required strength. The second approach is usually a much more fruitful one, but may involve iterated design of a product, for example a trussed beam, starting with several candidate materials and taking account of their individual mechanical properties before a true decision can be reached. This process can be simplified by using computer aids.

The possible ways by which a metal artifact may be produced always looms large as a factor in the designer's mind. Table 4.2 lists common production routes, and Table 4.3 expands the characteristics of some of them. In the context of an industrial company producing an article, the existing production facilities must be exploited; seldom will the designer alone be in a position to make a product design decision demanding entirely new manufacturing machinery. Producibility is related to material costs, since there are inevitably many direct, and even more indirect, sources of cost revolving round a manufacturing facility. Sometimes the decision to choose a material can depend not so much on its basic raw cost per unit volume, but on the fact that it happens to be offered in a suitable range of stock sizes, shapes and surface finishes, so that machining to final shape is minimised. Furthermore, some materials may incur additional costs because, to achieve the required design strength, heat treatments such as precipitation hardening or annealing might be necessary. To avoid treatment, the designer may opt for a product assembled from individual components by welding. It must then be remembered that welding is a heat-generating process which may give rise to undesirable (or, more rarely, desirable) metallurgical changes during manufacture.

Mention was made earlier of the influence of temperature resistance on the strength of metals. Due consideration must be given to the total service performance requirements to be met by the product. These may, on the one hand, invoke a mechanism working at a reasonably elevated temperature, such that there is danger of the metals used losing a significant proportion of their strength. On the other hand, during a shut-down period, a product may be exposed in a static condition to low temperatures at which point the materials may then become so brittle as to pose a severe risk from the effects of accidental impact or other damage. Whatever the temperature aspects of performance, the problem should be recognised and appropriate questions posed to the metal manufacturer or supplier.

Another feature common to most metals is the need for surface coating or protection. Corrosion of most ferrous metals is a well-known phenomenon and can involve expensive and tedious activities such as the continuous repainting of the Forth Railway Bridge. Some metals, unlike iron, may corrode in such a way such that its corrosion products are not more bulky than the basis metal, so that a surface film with good adhesion can be formed to assist in substrate protection. This happens particularly with stainless steels and some aluminium alloys, but is also seen in other metals such as copper. A distinction must be made between the type of finish needed to offer a technological function such as protection against corrosion or provision of a wear-resistant surface, and the

Table 4.2. *The most important production routes for metals*

Manufacturing methods	Aluminium	Copper	Heat and corrosion resistant alloys	Iron	Lead	Magnesium	Nickel	Precast metals	Refractory metals	Steel	Tin	Titanium	Zinc
Sand casting (green sand)	X	X	X	X	X	X	X			X	X		
Sand casting (dry sand)	X	X	X	X	X	X	X			X	X		
Shell mould casting	X	X	X	X			X			X			
Full mould casting	X	X	X	X			X			X			
Permanent casting	X	X		X	X	X				X	X		
Die casting	X	X			X	X					X		X
Plaster mould casting	X	X											X
Ceramic mould casting	X	X	X	X			X	X		X			
Investment casting	X	X	X		X		X			X			
Centrifugal casting	X	X	X	X		X	X			X			
Open die forging	X	X	X			X			X	X		X	
Closed die forging	X	X	X			X			X	X		X	
Upset forging	X	X	X			X	X			X		X	
Cold heading	X	X	X		X	X	X			X			
Impact extrusion	X	X			X	X						X	X
Blanking	X	X	X			X	X	X	X	X	X	X	X
Stamping	X	X	X			X	X	X	X	X		X	X
Drawing	X	X	X			X	X	X	X	X		X	X
Explosive forming	X	X						X					
Electromagnetic forming	X	X		X				X					
Electo hydraulic forming	X	X		X				X					
Spinning	X	X	X			X	X	X		X		X	X
Powder metallurgy	X	X					X	X	X	X			
Electroforming	X	X						X		X	X		
Extrusion	X	X			X	X	X		X	X	X	X	X
Roll forming	X	X			X	X	X			X		X	
Continuous casting	X	X	X	X	X					X			X

After N.C.W. Judd 1984

Table 4.3. *Some characteristics of metal production processes*

Process	Advantages	Limitations
Sand casting (green sand)	Low tooling costs Almost no limit on size or shape of part	Close tolerances difficult to achieve Rough surface finish Long thin projection not possible
Sand casting (dry sand)	As with green sand casting except that long thin projections may be formed	Restricted in smaller parts than is the case with green sand casting
Shell mould casting	Relatively high production rates Smooth surfaces Uniform grain structure	Patterns, equipment and mould materials are expensive
Full mould casting	Similar to green and dry sand moulding, but no draft and no flash	Patterns are expendable
Permanent mould casting	High production rates Good surface finish Mould re-usable	Mould casts are high Cannot be used with high melting point metals. Casting intricacy limited
Die casting	High production rates Good surface finish	Die costs are high Restricted to small parts
Plaster mould casting	Can produce intricate castings of low porosity with smooth surfaces	Restricted to small castings
Ceramic mould casting	Can produce castings of intricate configuration to close tolerances	Restricted to small castings
Investment casting	Can produce castings of intricate configuration with good surface finishes to close tolerances without flash	Expensive patterns Labour costs high
Centrifugal casting	For the production of large cylindrical parts	Restrictions on size of castings Shape of casting limited Expensive equipment

Process	Advantages	Disadvantages
Open die forging	Simple inexpensive tools Economic for small quantities	Slow production rates High degree of skill required Limited to simple shape Difficult to hold close tolerances
Closed die forging	High production rates Good reproduceability	High tool costs Machining often necessary
Upset forging	High production rates	Restricted to cylindrical shapes of limited size
Cold heading	Can be used for fairly intricate parts High production rates Good surface finish Low scrap loss	High tooling cost Restricted shapes and limited sizes Residual stresses in article
Impact extrusion	High production rates Low tool cost Good surface finish Low scrap loss	Tubular shapes only Part size and length to diameter ratio limited
Blanking	High production rates Good surface finish Variety of shapes and sizes	High tool costs Limited to thin sections Finishing often required Often involves high materials waste
Stamping	As with blanking	As with blanking
Drawing	As with blanking Low tooling costs Suitable for large parts	As with blanking Low production rates Skilled labour required
Explosive forming		Usually limited to symmetrical shapes Equipment cost high Configurations limited
Electromagnetic forming	Material can be worked in hardened conditions	Production rate low Skilled labour required
Electrohydraulic forming	Can be used for articles too large for conventional machinery	Specialised equipment required

Table 4.3. (cont.)

Process	Advantages	Limitations
Spinning	Low cost Good surface finish	Low production rate Low tooling costs Skilled labour required Restricted to symmetrical shapes of limited thickness
Powder metallurgy	High production rates Close tolerances Density and porosity can be controlled Low scrap losses	Limited shapes and sizes Not practical for small quantities
Electroforming	Can be used for large and small articles of intricacy Good surface finish	Low production rate Skilled labour required Limitations on configurations
Extrusion	Can be used to produce a variety of complex shapes	Restriction on size of article which must have uniform cross section Close tolerances difficult to achieve
Roll forming	High production rate Good surface finish	Large production runs necessary for economy Tooling costs high Parts must be of uniform cross section
Continuous casting	High production rate Low cost Can be used with alloys not easily worked	

After N.C.W. Judd 1984

type offering the aesthetic requirements of providing an adequate colour, degree of gloss or uniform matte appearance which will be retained throughout service life. These aspects of corrosion protection systems and appearance are developed in Chapter 9.

4.2   Characteristics of specific metals

Since the purpose of this book is to introduce the designer to materials technology in general, an exhaustive discussion of metals cannot be offered, not only because of space limitations but because so many excellent texts on metals already exist, for example Smithells (1983). It must suffice to provide a brief introduction to some of the more important commodity metals on which are based the range of alloys with which the designer will mostly have to deal. These comprise iron, aluminium, magnesium, zinc, copper and nickel. A brief description is provided merely to offer a 'flavour' of their likely service performance and of the way in which they can offer multiple choices to the product designer. A further purpose of this section is to highlight some recent developments in these metals to show that materials technology is still very much a living and burgeoning discipline. The selection of new advances is not meant to be exhaustive, and is clearly only illustrative of the potential which metals offer as a result of developments yet to come.

4.2.1   *Iron and its alloys*

Ferrous metals in practical form may consist very largely of iron with minor amounts of other elements such as carbon, or they may consist of alloys of iron with significant proportions of other metals. In general, ferrous metals are less expensive than other metals which derive from less plentiful ores or which require more expensive extraction processes. As befits a metal with a venerable history of service to mankind, many forms and varieties are available to the designer.

Cast iron, containing relatively large proportions of carbon in coarsely granular form, is known variously as grey cast iron, ductile cast iron, or malleable cast iron. In all cases it is shaped by a casting process, either into ingots for subsequent working or in moulds approximating the final shape, which may then be machined to finish it. Examples of use include lathe beds and pump housings. Cast steels, which can include significant amounts of alloying elements or which may comprise largely iron with a small amount of carbon, are available in stock shapes as sheet, plate, strip, rod and bar, and also are made from ingots into forgings.

Alloy steels can be categorised as those with high strength, or those which are stainless (and also resistant to heat and corrosion). The latter sub-divide into austenitic, ferritic, martensitic types, some of which are age-hardenable. In

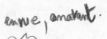

these days of the jet engine, super-alloys of these steels have come to the fore as major components for the engines' hottest parts; when containing significant amounts of other metals such as nickel, they are chosen where total reliability is required along with low creep and very high operating temperatures. Finally, steel, or rather iron-alloy articles, can be processed as metal powders pressed into a mould and subsequently heat-treated to sinter the compact into a part which should, in the best practice, closely approximate to the final required shape. More will be said of powder metallurgy later in this chapter.

Low carbon steel sheet is normally chosen when significant bending operations are required to fabricate a product, and where too a good surface is required to accept paint or electroplated finishes. For deep drawing or severe forming operations, other varieties of the material are available. Low carbon steel stock is available in free cutting form for general machining, and to provide a workpiece with a surface suitable for subsequent electrodeposition of finishes. Hot rolled low carbon steel flat bars are available for constructional applications where appearance is not of great importance, while cold-drawn bars are chosen for constructional use where a presentable surface finish is required. High carbon steel can be obtained as annealed strip for making flat springs, subsequently heat treated to realise optimum mechanical properties. High carbon steel strip can also be obtained in a hardened and tempered form for which forming operations are limited; the material is usually used as purchased. High carbon steel wire is available for making round wire springs, and the stainless steel equivalent is intended for resistance against various acids, salt solutions and gases.

Stainless steels are a comprehensive family of alloys with an enormous range of properties and hence of applications. The popular image of such steels are the '18/8' and '13 Cr' types, but the palette is very much wider than this. The next paragraph provides some detail on a restricted range of these steels, to illustrate what information is available. The designer is strongly urged to remember that expert advice, as from British Steel Corporation Advisory Service, Rotherham, is only a telephone call away.

Stainless steel sheet in the austenitic form has good resistance against various acids, salt solutions and gases, but may suffer corrosion if in contact with non-ferrous metals. Stainless steel strip in the ferritic or martensitic condition, unlike the austenitic, is magnetic. The recent development of super-plastic stainless steel offers the possibility to manufacture high quality, deep drawn sections at reasonable cost. Normal stainless steel, if to be deep drawn, requires a multiplicity of small drawing cycles to enable it to be fabricated. Microduplex stainless steels are characterised by a combination of high strength and good corrosion resistance, are well-proven in the chemical industry, and become super-plastic at temperatures of 1000 °C. In this condition they can be stretched up to more than 200 %, compared with normal stainless steel with an elongation value of only about 30 %. Knight (1982) discusses how super-plasticity permits low forming pressures to be used even when excellent shape reproduction is required from a die. The forming technique involves applying compressed air to one side of a hot sheet laid in the die, so that complex pressings, with depths up to 350 mm, can be made from sheet with thicknesses up to 2 mm and with areas up to 0.6 m^2. These microduplex steels recover their normal properties below 800 °C, so there are few constraints on normal service applications.

4.2.2   *Aluminium*

Aluminium is a silvery white metal characterised by lightness, lack of toxicity and ease of forming. Among common metals it stands second in the scale of malleability and sixth in ductility. Because of the high energy requirement for winning aluminium from its ores, this metal's use pattern is dominated by scrap recovery and recycling, which helps to lower the cost to no more than perhaps six times that of iron.

Although pure aluminium offers the great advantage of being very resistant to atmospheric corrosion it is too weak mechanically for significant structural use. For structural applications it must be strengthened by alloying with silicon, magnesium, manganese, zinc, copper, or combinations of these. Alloys can be divided into two broad classes; those which are cast, and those which are mechanically worked or wrought, the latter constituting the major use for structural applications. Casting alloys usually contain 8 – 12 % of silicon with small amounts of other elements, and are used for pressure or gravity die-casting, or for sand casting. Although these alloys have their various strong or weak points, it must be remembered that the high proportion of alloying element tends to reduce resistance to corrosion in some circumstances, particularly with respect to sea water. Castings generally have lower strength than wrought products, but can be heat treated to increase some properties to an acceptable value.

Plain sheet, profiled sheet, and strip and foil are produced from wrought alloys by rolling processes, while bars, sections and tube are produced by extrusion from cast billets. Cold working of some alloys increases strength and reduces ductility, a property which may be exploited if the geometry of a final product requires only a limited amount of rolling, drawing, pressing or stamping. Certain alloys in this category are suitable for precipitation hardening heat treatment which results in the internal deposition of alloy phases able to limit crack growth.

For many purposes, aluminium can be used as it comes from the production process, being protected from mild corrosion by a weather-resistant and transparent film of oxide. However, should this be required to withstand hard abrasion, or present a more uniform appearance, a thicker film of high density oxide can be built up on the surface by an electrolytic anodising process. From this point on, some other decorative finishes can be applied. When deep drawn configurations are needed, perhaps in small numbers for which the use of special press tools would be uneconomic, it is possible to employ super-plastic aluminium alloys which can be formed into contoured shapes by drape-or-vacuum-forming over a male mould, or by gentle pressing into a female mould, using air pressure or very inexpensive press tools.

A recent development is commercial introduction of a high strength aluminium – lithium alloy, 10% lighter and 10% stiffer than conventional alloys used in the aerospace industry. Increasing the lithium content up to 5% by weight reduces the density to about 2.4 and increases the Young's modulus from 67 GPa to 86 GPa. Application of this material in the aerospace industry is discussed by Knight (1983*a*), who suggests that production bottlenecks in making the basic alloy could retard its large-scale introduction. Despite the reactivity of lithium as a pure metal, once alloyed with aluminium the end result is manageable by conventional processing methods.

4.2.3   *Magnesium*

Magnesium is one of the most abundant metals at the earth's surface, being commercially available from inland salt lakes and salt deposits, and from the sea. It is lighter than aluminium, silvery white and is finding its way gradually into the market-place for light structural applications, in addition to its traditional uses as alloys with aluminium, with which it compares favourably on a strength-to-weight basis. As with aluminium, magnesium alloys are available both for casting and for forging. The more common cast alloys contain aluminium, zinc and manganese, and specialist alloys contain rare earth metals. Certain magnesium alloys are particularly suited for die casting at high production rates, with the advantage over aluminium of reduced attack on the steel moulds.

Wrought alloys are chosen for vehicle parts, electronic equipment, office machines and for many other applications. Like castings, they all have poor intrinsic corrosion resistance and so must either be used in uncontaminated atmospheres or be coated with a plastic resin or other specialist finishing system. Generally, magnesium alloys are not suitable for cold working, but are capable of deformation at around 300 °C, so extrusion is the most common method of hot working the metal. Because of moderate temperature weakness, the upper service temperature limit is about 50 °C.

4.2.4   *Zinc*

This element, widely found throughout the world in association with other metals, is seldom used on its own except as flashing sheet in building work, where it is chosen for its impervious, tenacious and protective grey oxide film, which eliminates the need for additional surface finishing. Zinc is offered commercially as cast slabs, which may be used in smaller pieces as sacrificial anodes for corrosion protection of structures such as bridges and ships. However, for structural use it is always formulated with aluminium, copper, magnesium, iron or other metals to provide a range of wrought or cast alloys.

Casting alloys are excellent for reproducing complicated shapes at a low casting temperature, offering a number of production advantages. In many ways zinc die-castings compare favourably in acceptability with plastic mouldings, except that the latter require less finishing, and are lighter. Zinc castings, like wrought zinc alloys, are corroded by acids and by strongly alkaline reagents, and in the presence of salt generate a white powdery surface deposit which is not adherent and thus not protective in the long term of the basis metal.

Wrought zinc alloys are available as plate, strip, sheet, extruded rod, drawn rod and wire. From these standard shapes items are produced by deep drawing, impact extrusion, bending, roll forming, stamping, swageing and other processes. Usually, zinc alloys are not selected for highly stressed structural applications because they have a rather low specific stiffness compared with other cheap metals, and they have a strong tendency to creep under continued stress.

Despite the rather soft surface of pure zinc, a new group of high-performance zinc alloys with high aluminium content has recently been claimed to offer excellent bearing characteristics; these might be considered for low-speed applications as a replacement for bronzes and leaded bronzes. Calayag & Ferres (1982) discuss these white bronzes, which have been known since the early twentieth century but which could not, until recently, be exploited until problems of corrosion and ageing were overcome by control of composition. The advantages of these materials for bearings include not only technical factors such as low frictional coefficient, absence of shaft seizure and good embedding properties, but commercial and ecological ones such as low prices, energy savings and pollution-free processing.

4.2.5   *Copper*

Copper as a native metal has been known from antiquity, and hence has a long history of human efforts to purify, alloy, fabricate and harden it.

Although its most valuable single property is probably the high electrical conductivity, copper alloys can be used for many structural, decorative and technical purposes. Unlike stainless steels and some aluminium alloys, copper alloys do not rely upon a film-forming tendency to protect the surface from environmental corrosion. Brasses, bronzes, aluminium bronzes, gun-metals, cupro-nickels and many other copper alloys are accordingly used extensively for marine applications. Wrought alloys tend to contain very minor (almost accidental) proportions of oxygen, phosphorus, or other materials, but are deliberately doped with small amounts of magnesium, beryllium, chromium, zinc, for specific technical purposes. In general, very pure copper has high resistance to a variety of atmospheres, although it will turn green when exposed to sulphur compounds, a property exploited in roofing applications. Wrought alloys, commercially available as sheet, strip, bar, rod, plate, wire, tube and extruded shapes, are used for electrical conductors, heat conducting gaskets, architectural uses, and for machinery and electric motor parts. The more heavily wrought alloys, which are available in preshaped forms as well as ingot, are used for heavier engineering applications of a similar kind, particularly in the electrical engineering industry.

Brass is an alloy of copper and zinc, but lead comprises a main alloying element also. Brasses are stronger and cheaper than other copper alloys, and are used for constructional purposes. Some alloys can be strengthened by precipitation hardening treatments, operating by the same mechanisms which apply to aluminium and ferrous alloys. Alternatively, solid solution strengthening is applied, in which alloying elements enter the crystals of copper to produce a homogeneous alloy structure.

Cast alloys, with only minor additions from a wide range of elements, are available for electrical conductors and contacts, for machinery parts, cams, bushings, bearings and pump parts, as well as for valves, flanges, pipe fittings and various marine parts. Very large castings can be made of which physical and other properties of the numerous available types are well documented in Materials Engineering (1982).

A fairly recent development in copper alloys is oxide-dispersion-strengthened copper available from the USA. McIntyre, R.D. (1982). This is made by a powder metallurgy process in which inorganic oxides are dispersed in copper powder, pressed and sintered. This yields a product with high strength and electrical conductivity able to withstand operation at elevated temperatures for long periods without recrystallisation. The latter phenomenon in other coppers can significantly reduce both hardness and mechanical strength. Anon. (1981*b*).

Michael & Hart (1980) report the development of a range of copper, zinc, aluminium alloys (SME brass in the UK) which demonstrate a shape-memory effect. These alloys in the low-temperature martensitic condition can be deformed, and retain the deformation at room temperature. When heat is applied, the original shape returns, strains typically of $6-8$ % being completely recovered. Michael & Hart describe so many intriguing properties that a designer could easily be induced to consider applications such as a thermostatic clutch or ventilator controls (to mention only two).

4.2.6 *Nickel*

Compared with the metals treated in earlier sections, nickel is considerably less available worldwide but compensates by being pervasive in metallurgy. Its relative rarity can be seen from the fact that figures for annual production and consumption are quoted in millions of pounds weight rather than in tonnes. The material is light-grey, tough, ductile and slightly magnetic. It is seldom used in massive form on its own, except for electroplating anodes, but when alloyed with a minor proportion of beryllium, it is employed as wrought or cast alloys for various jet engine parts, for springs, switches, diaphragms and other technical applications. Wrought nickel alloys with chromium and iron are used for electrical resistance applications, while alloys with copper, molybdenum, manganese and silicon are used for magnetic applications in relays, magnetic shields and communications equipment. The Monel range of copper nickel alloys is intended for applications requiring high corrosion resistance such as valves, pump parts, propeller shafts, heat exchangers, springs and similar applications; a range of chemically-resistant alloys under the generic name Hastalloy is available for the chemical process industries; the high temperature alloys bear trade names such as Inconel and Incolloy. Some important production routes for these metals are shown in Tables 4.2 and 4.3.

4.3 Welding

The techniques of welding are covered in rather more depth in Chapter 8 where the performance of metals in service is discussed. However, the designer should always be aware of the metallurgical effects, sometimes deleterious, that can occur if materials to be welded are not chosen adequately. For example, Smallen (1981) in the context of hermetically enclosing hybrid electronic microcircuits, reported that although a flat package was made from a controlled expansion metal (Kovar) chosen to match the thermal expansion of contiguous glass seals, a cracking problem occurred after welding with an alloy of silver and copper because a ternary alloy formed with some gold present, this alloy being subject to cracking. Furthermore, the welding process can leave high residual tensile stresses which will drastically reduce fatigue resistance. Heat treatment to reduce these stresses is not always possible. The message to the designer is that if he suspects that any metallurgical interactions might occur, or even if he is merely cynical enough to look for trouble for its own sake, he would do well to discuss the matter with an expert and, furthermore, should check that the expert is as skilled as he claims to be.

4.4   Machining

Reference has been made earlier in this chapter to machining processes for metal parts. This is a subject for which the designer should have

Table 4.4. *Taylor (machineability) speeds for metals*

	Metal	Brinell hardness	Taylor speed
Ferrous			
AISI B1113		179−229	135
	Standard malleable iron	110−145	120
AISI B1112	Bessemer free cutting steel		100
AISI B1111		179−229	95
AISI C1109	Low carbon desulphurised steel	137−166	85
	Soft cast iron	160−193	80
AISI C1132	Medium carbon desulphurised steel	187−229	75
	Medium cast iron	193−220	65
AISI A4815	Low carbon wrought alloy steel	187−229	50
	Hard cast iron	220−240	50
	Stainless 18−8 free cutting	179−212	45
	Tool steel (low tungsten, chromium and carbon)	200−218	30
	High speed steel	200−218	30
Non−ferrous			
Wrought magnesium with Al, Mn, Zn		58	500−2000
Cast magnesium		58	500−2000
Aluminium alloys		20−95	300−2000
Brass, leaded, cold drawn		100	200−400
Zinc		80−90	200
Phosphor bronze, leaded		60−100(Rockwell)	100
Yellow brass		—	80
Cast copper		30	70
Manganese bronze (Cu, Zn, Sn, Mn, P)		55−210	60
Aluminium bronze (Cu, Al, Fe, Sn, cast)		140−160	60
Gun metal (Cu, Sn, Zn)		65	60
Rolled copper		80	60
Nickel		110−150	55
Cast Monel (Ni, Cu, Al)		175−250	40

some 'feel' before specifying materials which must be machined as part of the product production process. This is not to say that he must rush to the cutting speed tables when considering each new project, but he should have some idea of differences in machineability between materials, and the relation they have to surface hardness, so that unsuitable materials can be avoided from the beginning.

Machineability can be defined as an index of fast removal of material to a satisfactory finish without the necessity for excessive cutting tool repair. It can be assessed by several parameters, such as measurement of power consumed by the machine tool, rate of penetration of a workpiece by a standard drill under constant pressure, and determination of maximum temperature generated under constant working rate conditions.

The processes of machining include turning, milling, drilling, boring, tapping, planing and sawing, which can be applied to a wide variety of materials. Non-traditional machining methods include abrasive flow, abrasive jet, hydrodynamic, thermally assisted, ultrasonic, and water-jet machining. Other processes, also non-traditional but non-invasive regarding tooling, include electrochemical, electric discharge, plasma beam and chemical machining.

Trace impurities in, or the deliberate addition of certain elements to metals and alloys can affect machineability as well as other properties such as corrosion resistance. Thus for steels, addition of an appropriate concentration of carbon, sulphur, selenium, phosphorus, lead, zirconium or titanium, separately or in combinations, can tend to increase the ease of machining; on the other hand levels of manganese, silicon, nickel, chromium, molybdenum, vanadium, aluminium, copper and tungsten, above a certain low threshold value, will tend to reduce machineability, although they do have beneficial effects in other directions. Apart from these effects, the microstructure of steels (and hence their machineability) is often changed by direct additive incorporation, so the phenomenon of element segregation, especially in small castings,

Table 4.5. *Relative power requirements for machining metals*

Metal	Percentage of power required
Magnesium alloys	100
Aluminium alloys	55
Brass	45
Cast iron	30
Mild steel	20
Nickel alloys	10

leads to variations in machineability across a workpiece. Incorporation of trace impurities or deliberate additives in aluminium, magnesium, copper, nickel and zinc alloys can also affect machineability for good or ill.

Machining Data Handbook (1980) provides information, particularly on surface speeds, feed rates and preferred tool materials for a very wide range of materials identified as individual alloys, and on the common machining processes. Materials described include ferrous alloys, aluminium, magnesium,

Table 4.6. *Single point turning settings for machining materials with materials with high speed steel tools*

Material	Brinell hardness	For depth of cut (mm)	Surface speed (m/minute)	Feed rate (mm/revolution)
Carbon steel (free machining, wrought)	100–150	1	60	0.18
Medium carbon (wrought alloy)	375–425	1	18	0.13
High speed tool steel	225–275	4	18	0.40
Ferritic grey cast iron	120–150	1	56	0.18
Aluminium alloys	30–80	1	305	0.18
Magnesium alloys (wrought)	55–60	4	275	0.40
Copper alloys (wrought)	—	1	145	0.18
Monel (Ni,Cu,Al)	115–240	1	30	0.18
Zinc alloys (cast)	80–100	4	100	0.40
Lead–tin alloy	5–15	1	150	0.18
Tin alloy (Babbitt)	15–30	8	76	0.40
Machineable glass–ceramic	—	1	15	0.05
Thermoplastics (varies with type)	—	1	120–150	0.13
Thermoset resins (varies with type)	—	1	60–150	0.13

copper, nickel, zinc, lead, tin, glasses, ceramics, plastics, composites, plated materials and rubbers.

An old measure of machineability is the 'Taylor speed'. By definition, AISI B1112 Bessemer free-cutting steel has a Taylor speed rating of 100, being cold rolled or cold drawn, and machined with cutting fluid at a surface speed of 55 m min^{-1} as a reference sample. Table 4.4 shows Taylor speeds for some metals.

Another index of machineability, that of the relative power needed for machining a workpiece, is exemplified for some metals in Table 4.5. Examples are also provided in Table 4.6 of machining parameters necessary for processing some metals and non-metals, using turning with single-point, high-speed steel tools.

For planing operations, some representative figures are given in which carbide tools are used (Table 4.7); Table 4.8 provides figures for power hacksawing, and Table 4.9 supplies brief data for power band sawing.

Each metal has its own particular response to machining. For example, Stasko (1980) reports that deflection of the workpiece and metal build-up on

Table 4.7. *Settings for planing metals with carbide tools*

Material	For depth of cut (mm)	Surface speed (m/minute)	Feed rate (mm/stroke)
Free machining carbon steel (wrought)	2.5	90	2.05
Low carbon wrought steel	12.0	84	1.5
Wrought high speed tool steel	0.1	52	—
Aluminium alloys (cast)	2.5	90	3.2
Magnesium alloys (wrought)	12.0	90	2.3
Copper alloys (cast)	2.5	90	1.5
Nickel alloys (cast)	2.5	79	1.5

Table 4.8. *Settings for power hack sawing of metals*

Material	Material thickness (mm)	Tooth pitch (mm)	Feed rate (mm/stroke)
Aluminium alloys	6–18	4.0	160
Magnesium alloys	18–50	6.3	130
Copper alloys	6–18	2.5	40
Tin alloys (cast)	50	4.0	40

Table 4.9. *Power band sawing rates using a diamond coated band*

Material	Material thickness (mm)	Band speed (m/minute)	Diamond grain size
Glasses	12	915	80−100
Ceramics	12−25	610	80−100
Kevlar 49−epoxy resin	25−50	610	60−80
Fibre glass−epoxy resin	12−25	915	60−80

cutting tools are just two of the difficulties which aluminium offers the machinist. Although the metal is easy to cut and cutting tools can last a long time, this is definitely not easy machineability. Stasko discusses successful machining requirements for aluminium in general, but particularly the casting alloys, in terms of optimum tool angles and design, together with the various techniques for achieving a good surface finish.

4.5   Surface conversion

Although finishes and coatings as protective systems are dealt with in Chapter 9, a preview must be given in this chapter on metals because of their special needs. The designer may be familiar with case hardening a ferrous metal by immersion in molten sodium cyanide, but may not be so familiar with ion implantation, nitriding or metallising. The purpose of these and other treatments is to develop a hard, wear-resistant or tough surface skin when it is not practicable for technical or cost reasons to convert the whole mass of metal in this way, or to use a harder metal as a first choice. If one assumes that a surface skin can be extremely resistant to abrasion, but is also intrinsically brittle, then obviously it is an advantage that the substrate metal should be reasonably tough and supportive, needing to be modified only to the extent of a very thin surface layer to provide the performance characteristics required.

Brief details of many surface treatments for steels and for other metals are provided by Ross (1977). Ion implantation, pioneered by the Atomic Energy Research Establishment at Harwell, involves introducing by bombardment the desired atomic species into the surface of a material. The process significantly improves wear resistance and fatigue endurance of some metals such as steel, titanium, copper and electrodeposited chromium, the effect being explained in terms of the pinning of mobile dislocations in the treated volume, unlike oxide-dispersion strengthening which strengthens by pinning dislocations throughout the material. Ion implantation can be applied to finished parts which undergo no dimensional changes; thus it is suitable for application to machine tools.

5

Properties of ceramics

5.1. General characteristics

In the present context, we shall be using 'ceramics' and 'glass' as terms for ceramic materials intended for engineering applications. Although in some ways one might consider the structure of Portland cement to be allied to that of ceramics, such cement will not be dealt with at all except for later mention of a new development in which porosity has been significantly reduced.

Ceramics are usually based on both metallic and non-metallic elements joined by atomic bonds that are partly ionic and partly covalent. This type of bonding, different from that in other materials such as minerals, gives rise to the characteristic hardness, brittleness and heat resistance of ceramics. Ceramics comprise those inorganic materials (with the exception of metals and their alloys) which have been subjected to a high temperature processing stage to produce solid objects from the powdered starting materials. Ceramics are usually crystalline with single crystals or with dense polycrystalline masses, but there also exist composite materials with ceramic crystallites bonded into a glassy matrix, as well as the totally non-crystalline glasses. The general structure of these materials is seen in Figure 5.1.

In the crystalline materials, atoms are joined together in a repetitive pattern which follows through from one basic unit of molecular structure to the next, so that a regular array of atoms is seen on the larger scale. On the other hand, amorphous materials may have some short range order of atoms but over the larger physical dimensions in a material, there is no orderly arrangement of the atoms. Unlike metals, which have comparable tensile and compressive strengths, ceramics as a class are very strong in compression but usually weak in tension. Their stress − strain characteristics are such that only a modest strain is accepted before fracture occurs.

5.2　Varieties available

Ceramic materials are often binary compounds such as those of tungsten, molybdenum, vanadium, hafnium, zirconium, and tantalum with non-metals such as silicon, boron, nitrogen and carbon. Perhaps the best-known carbon compound is silicon carbide, long used in grinding wheel and abrasive technology. Latterly, silicon nitride has assumed some prominence as a potential material of construction for high temperature applications, including some parts of internal combustion engines. Some ceramics are based

Fig. 5.1. Idealised structures of crystals, glasses and ceramics. (Copyright: 1985 Standard Telephones and Cables plc.)

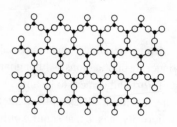

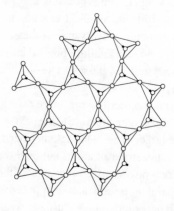

Crystalline material : short-range symmetry and long range order of the atomic structure

Non crystalline glass : short-range ordered structure but no long range order

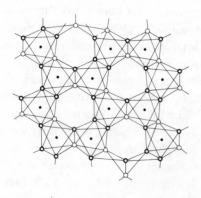

O Oxygen　　　● Silicon

Distorted silica crystals

● Aluminium　　O Oxygen　　◉ Hydroxyl

Part of typical aluminium silicate atomic structure

on sulphides, of which cerium sulphide has been used for manufacturing crucibles for metal melting. Other ceramics are made of simple oxides, of which silicon, aluminium, magnesium and zirconium are relatively well known. Some complex oxides of commercial importance include mullite, zircon and some high alumina and high magnesia fire-brick materials. As a generality for all such materials, it can be said that the purity and consistency of starting materials and refractory properties go hand in hand — the more pure, the better the performance usually achieved. Sometimes best refractoriness is achieved by close control of composition rather than absolute purity; 96% alumina is an example.

Another class of materials related to ceramics are composites involving ceramic fibres in a metal matrix, and those in which ceramic and metallic phases are randomly and intimately dispersed on a microscopic scale to comprise a cermet. For example, metal-based fibre-reinforced composites are used for aerospace hardware, particularly where high temperatures and thermal shock are to be met, whereas cermets are particularly hard materials intended for machining purposes. Cermets are also made from dispersed chips of ceramic (say, a cutting material) held in a bonding metal matrix.

Glasses are the most widely used non-crystalline ceramic, being available not only as the very well known soda lime and borosilicate glasses used for windows and containers in the home and in the laboratory, but also in a range of special formulations including those with high lead oxide content for optical purposes. Other formulations are based primarily on phosphates and fluorides, intended for slow dissolution in water to release contained solutes.

Many ceramics are produced in volume as standard-shaped items, of which refractory bricks, tubes, insulators, and thermocouple protection tubes are relatively well known. In addition, standard items are available for specialised uses such as capacitor dielectrics, hermetic seals and fibre-boards. Where possible, of course, the designer should select from standard stock catalogues, provided that his product has no unusual performance needs.

Also classified as ceramic, but not elaborated here, are some of the natural mineral silicates and various clays used for pottery and fine china-work. They will be discussed only incidentally in Section 5.4 dealing with fabrication processes.

Although until very recently all ceramics and glasses were made from powdered starting materials made to react and densify by high temperature processes, a sol-gel method has been developed in which reaction occurs at ambient temperature in water, being followed by a sintering and consolidation process which takes place at much lower temperatures.

A useful summary of the varieties of ceramic materials is given by Hamme (1979).

Although technically classed as ceramics by materials scientists, ferrite materials commonly used for magnets are not included in the present discussion, which is confined to those materials intended for engineering applications. For similar reasons, piezoelectric and optoelectric ceramics will not be discussed; Anon. (1981*e*) provides a useful reference to these materials.

5.3   Comparative properties

A full exposition of the properties of even a few commercially available ceramics would occupy an unreasonably large proportion of this book. Table 5.1 provides illustrative data on a few of the better known ceramics to show the breadth of properties which are available. Authors such as Newnham & Davies (1976), for example, should be consulted for more complete data. By far the most important properties of engineering ceramics lie in their combination of hardness, ability to withstand high temperatures and extreme lack of ductility. The latter, which leads to catastrophic failure in the presence of small surface cracks under applied stress, is responsible for the commonly observed brittleness, although it should be pointed out that not all modern engineering ceramics are brittle.

Ceramics are not readily shaped or worked after firing, except by very expensive grinding processes, usually using diamonds, or by laser machining, so there are some constraints on the exploitation of favourable properties. Accordingly, all the required functions and the shape and size of the piece to be finally obtained, must be catered for, beginning with the very first stages of processing. As a class they are characterised by very potent electrical properties, particularly in respect of electrical break-down strength and resistivity. The high surface hardness and resistance to abrasion which typify ceramics make them suitable materials for some electrical applications such as stand-off insulators for electrical power transmission. However, traumas such as thermal shock, impact and some chemical influences can damage the materials irreparably. Those ceramics comprising binary oxides and, to a lesser extent, the borides, carbides and nitrides, offer high chemical stability over wide ranges of temperature, being resistant to further oxidation, and often to reduction by metals as well. Much of the weakness of crystalline ceramics when in service can be ascribed to the presence of foreign impurity elements, or to defects in the crystal structure. Some impurities can, for example, lower the service temperature significantly; on the other hand such materials are often added deliberately to reduce sintering temperature, and thus manufacturing cost, even though the penalty is a reduced service performance.

Many ceramic substances are optically transparent either as single crystal or in vitreous forms; the transparency of glass is its most characteristic and useful

Table 5.1. *Properties of some ceramics*

| Material | Specific gravity | Tensile strength (MN/m²) | Hardness | | Compressive strength (GN/m²) | Coefficient of linear expansion ($\times 10^{-6}$ per °C) | Softening temperature (°C) | Young's Modulus (GN/m²) | Thermal Conductivity (W/m K) | Three-point transverse breaking strength (MN/m²) |
			Moh's scale	Vickers (GN/m²)						
Electrical porcelain	2.4	17–56	6.5		0.5–0.8	4.5–6.0		70–100	2.0	10–150
Mullite porcelain ($Al_6Si_2O_{13}$)	2.6		7.5		9.0	6.0	1800	145		
Fused silica	2.2	7		7.5		0.5	1580	70	1.5	
Borosilicate glass	2.4	35–140		7.0		3.4	920	60–80	1.0	60
Soda glass	2.49	45		5.5		8.5	825	60–80	1.0	50
Alumina (sintered)	3.6–3.7	90–270	9.0	25.0	2.5	7.2	1600	260–360	8.0	300
Aluminium nitride	3.25		8.5		21.0					
Boron nitride (hot sintered)	1.9	2.0		3.0	5 – 5	3000 (MPt)	48	29.0	70	
Silicon nitride (hot pressed)	3.2	105	9.0	25.0	3.0	3.2	1700	310	19.0	600
Tungsten carbide	15.6	350	9.0	18.0	4.4	4.9	2770	580	85.0	

feature. Dispersion strengthened ceramics lose this transparency because of the optical scattering effect of the small grains of disperse phase which are present.

Ceramics in general are strong in compression but weaker by factors of ten or more under transverse bending or in tension, the reasons for tensile failure by cracking having briefly been mentioned already.

Thermal expansion characteristics are perhaps worth a little explanation. If a polycrystalline ceramic is made up of cubic single crystals, the thermal expansion coefficient is the same in all directions. However, where the single crystals are anisometric (that is, all other types than cubic), thermal expansion will vary with the different crystallographic axes. Any method of fabricating the ceramic which effectively aligns these crystals will produce predictable consequences for the thermal expansion. When anisometric materials are processed into a polycrystalline ceramic body, the net thermal expansion may be very low owing to cancelling out effects. Materials undergoing very little dimensional change as a function of changing temperatures are, of course, able to withstand extreme thermal cycling and thermal shock traumas without fracturing; of these, silicon nitride is a noteworthy example, although fused quartz is perhaps more familiar to the non-specialist.

Creep, appreciable at high temperatures in some ceramics when structural loads are present, may be a life-limiting factor, although it is not normally critical for non-specialist low temperature work. On the other hand, static fatigue is a parameter which must concern the designer at all temperatures. This process, also known as stress rupture, involves sub-critical crack growth at stresses lower than those needed for brittle fracture. Initial flaws in the material slowly grow under the unremitting stress to such a size that instantaneous fracture can occur under a very slight additional load. Stress rupture tests are used for predicting the life of ceramic materials, but data are not easy to interpret because of the large scatter of results seen in any set of replicates. It is best, in considering stress rupture, to take specialist advice from the ceramic manufacturer.

Little has been said so far about the performance of glasses. These are perhaps better understood by the average engineer, and the literature is replete with physical property information. Miska (1980) states that, as with all ceramics, the material is inherently very strong, reaching 2700 MPa of tensile strength for glass fibres protected from all adverse influences. However, in the real world, as do other ceramics, glass fails without exception in tension because of flaws on the surface; stress concentrates at flaw tips, stress which glass cannot yield to accommodate. However, chemically strengthened glasses overcome this defect to a large extent by providing a surface coating which fills the flaws and prevents their uninterrupted growth. Although Miska's applications included high-strength wind-shields, tape reels and lens systems, a familiar domestic example is the humble milk bottle which has a surface chemically toughened by vapour treatment with a titanium compound.

Ceramics still have to live down their early reputation for brittleness and lack of impact resistance, even though modern developments of surface and bulk toughening of ceramics are well established. A single example will suffice. Davidge (1983) reports enhancements of strength and toughness in ceramic matrices such as alumina by incorporation, by heat treatment, of pure zirconia in the correct crystalline form. The process produces zirconia particle sizes with diameters of one micrometre or less, which under impact result in a partly compensating compressive stress being set up just ahead of a crack tip. This makes crack propagation much more difficult and thus both toughness and strength are increased.

Finally the designer should remember that ceramics, even more than other materials, show a strength property which is not unique to the material but is more likely to be unique to the individual test sample which has been evaluated. The measurements of any strength parameter on a number of test pieces will show a statistical scatter, which might reach 15% or more between the set. This means that extra attention must be paid to permissible safety factors, bearing in mind that the test to destruction of any representative sample from a batch of parts will not necessarily give a usable figure.

5.4 Data on some specific ceramics

Ceramics are manufactured from mixtures of starting compounds such as metal oxides and simple salts. A useful aide memoire on those used for industrial and electronic ceramics, particularly of later generation developments, is available from time to time in Ceramics Industry (1981). This lists the major materials used by the industry, describing physical properties, their method of preparation, applications and difficulties in use. Alumina and quartz are very well known, and their property data are readily available in texts. The next four sections deal with perhaps less familiar ceramics.

Sialon

A recent British development with very great potential for a wide range of applications is the material generically known as sialon, being a compound of silicon, aluminium, oxygen and nitrogen. Developed by workers at the Lucas Research Centre, the material is now on sale under the trade name SYALON. The material is made by replacing some of the silicon atoms in the crystal lattice of silicon nitride by aluminium atoms, while nitrogen atoms are displaced by oxygen. The result is a ceramic alloy which is sinterable like a ceramic and can be made fully dense (that is, with negligible porosity). The usual preforming processes can be used and firing leads to shrinkage which, being linear, is a great convenience for the designer. Applications for the material are as cutting tools for metal, with a potential in a wide range of industries exploiting the fact that the material retains its full room temperature strength up to 800 °C (unlike most metals) and is only derated by 25% when working at 1300 °C. Wright (1983) briefly discusses some of the properties, while Anon. (1983) in particular focuses on the use of the sialons for metal cutting.

Silicon nitride

Silicon nitride is an industrial ceramic with particularly high resistance to thermal shock, as shown by the fact that it can survive constant cycling between 1500 K and ambient temperature without losing strength. The production process starts with a high quality silicon powder which is consolidated by pressure and then sintered. The material is now machined with conventional tools or otherwise shaped into the required component size and geometry, and the parts are finally fired in nitrogen, converting the metallic silicon into silicon nitride by what is called a reaction bonding process. This treatment strengthens the part but does not produce any distortion or shrinkage. The material has excellent compressive strength and a transverse rupture strength comparable to mild steel, with the additional advantage of having the low density of aluminium.

Silicon carbide

Silicon carbide is a ceramic with a venerable history from 1891 and is familiar to every engineer as a major component in abrasive grinding wheels and blocks. It is available now in very large pieces suitable for use as high temperature structural components in rocket nozzles, refractories and hot pressing dies. The material has excellent thermal shock resistivity, coupled with a high thermal conductivity but low electrical conductivity. The properties of the fired ceramic show a wide range of values, particularly for compressive strength, due to the influence of firing aids and other deliberately incorporated components.

Zirconium oxide

Zirconium oxide has already been mentioned in the context of toughening other ceramics. However, it does form the basis for ceramics in its own right, and produces high performance materials with high resistance to wear and to corrosion. Usually these materials contain yttrium oxide as a sintering aid. After pressing and firing they produce fine grained, high density materials which are used in the electronics industry for sensing and fuel cells, and in aerospace applications. The phase transformation between tetragonal and monoclinic in zirconia preclude the use of the unstabilised material in bulk form for refractory bricks, but it is used in finally divided form to stabilise other materials, especially to enhance thermal shock resistance. When applied in a thin layer to metal turbine blades, gas turbine engines can be run at temperatures up to 200 °C above normal because of the high thermal insulation characteristic of the zirconia. The material finds its way into high-temperature fuel cells as an electrolyte. Zirconia in the form of tubes is also used as a heating element because it conducts sufficient electricity when preheated above 1000 °C to maintain a furnace temperature around 1600 °C. This material has advantages over a metallic element, not only in terms of the upper temperature available (above 2000 °C for short times) but because there is no oxidative degradation of the material. Stevens (1983) provides a useful summary of zirconium oxide ceramics and properties.

5.5 New developments in ceramics and glass

Despite the very long tradition of ceramics and glass technology, new developments of these materials are continually being reported. The following few paragraphs provide a flavour of what has recently become available.

Ceramic-fibre paper has now been produced in a range of thicknesses and porosities to act as an easily shaped heat insulator for operation up to 1400 °C (and sometimes even higher) up to the fibre softening temperature of 1700 °C. The material can be cut to shape, folded, wrapped and fitted easily into small spaces, without the health hazard associated with asbestos which previously filled these roles. The materials are soft to the touch and compressible, being suitable for use as insulation in domestic appliances such as toasters, as a replacement for asbestos in automotive exhaust silencers, for general insulation and use in the ceramics and glass industry, and for many other purposes.

A comparable material but this time made of non-woven glass fibres, is also available to replace asbestos where a flexible heat insulation material is needed. However, a recent development by ICI Mond Division in the UK involves modifying the surface of each glass fibre with an inorganic material which is comparable to vermiculite. This not only binds the glass together to retain its structural integrity, but insulates the glass from externally applied temperatures so that the material shows only 2% shrinkage at 1000 °C, which is much higher than one would normally expect a glass fibre to operate at. The materials can be used for fire- or flame-resistant facings for organic polymers and foams, and for interlayers and separators in domestic appliances and high temperature machines.

The world's first light transmitting ceramic material was recently developed in Japan by the Tokuyama Soda Company. The basis of the material is an ultra pure aluminium nitride powder with an oxygen content of about 1%, much lower than that of previously available aluminium nitrides. A reduction also in the level of cationic impurities in the material allowed sintering at atmospheric pressure so that the material is suitable for mass production. Property features include a thermal conductivity two to three times that of previous aluminium nitride ceramics, with a low dielectric loss and a coefficient of thermal expansion much less than that of alumina, making the material suitable for high-temperature windows resistant to thermal shock.

Cement and concrete have been available to the world for many years. A recent development in this area, which takes it from the outdoor industry into more refined engineering applications, is a British improvement on the material, by means of which the entrained moisture and air in cement and concrete mixes is effectively removed. In conventional concrete, the flexural strength is low, and holes caused by water evaporation account for 30% of the total volume. Recent work with additives carried out by Imperial Chemical Industries is aimed at eliminating this porosity by reducing the water content to a minimum, packing the powder closer and improving the adhesive quality of the cement by including a small volume of polymer to act as a lubricant. The result is a cement with similar bending and strength properties to aluminium, with an impact resistance which, although less than for metals, is similar to that of acrylic plastic resins. The material can be moulded like a plastic, machined and drilled like metals and also tapped, threaded and riveted. The material has been fabricated into coil springs with acceptable properties.

Lastly, mention must be made of the new sol-gel process for making glasses. Although ceramics and glasses are normally manufactured by dry mixing of starting powders, which are then chemically reacted at high temperatures, it is possible to produce these materials at room temperature from aqueous solutions, by mixing chemicals which react together to form compounds with a narrow distribution of particle sizes and a small mean size. These compounds, after drying, can be processed by conventional ceramic or glass techniques and allow sintering to high density. The sol-gel technique demands lower firing temperatures than are usually necessary to make solid bodies, and allows much more freedom in choosing the chemical composition of ceramics than used to be the case. Recently, Woodhead and Segal (1984) reviewed the essential features and advantages of the process as applied to the preparation of a range of ceramic materials. The same type of process applied to glasses has been discussed by Klein (1982). Glasses can now be made not only with high purity raw materials but the process can be used to produce materials which are amorphous, of high hydroxyl content or of high surface area, which means that the structure can be engineered on the molecular scale. An easily understandable application is that of glazing porous tiles, using a sol-gel approach to infiltrate a bonding material into fibrous materials, which promote fusion at the fibre contact points. This has been used for making some of the heat-resistant tiles on the US space shuttle.

5.6   Fabrication processes

A typical ceramic is made by mixing powders of controlled particle size, which are then compacted into an object of similar shape to the final component required, but with a larger volume, and held together only by the attractive forces of incorporated binders. This 'green state' part can then be gently machined to refine the shape if necessary; for example, holes may be bored in it. The part is then heated to initiate reactions which join the particles together into one cohesive whole. The finished component, which has shrunk during firing, may perhaps then be further finished by grinding significant surfaces.

The green compaction state can be avoided by hot pressing the powder mix, which effectively combines compaction and firing in one process.

Firing temperatures range from 700 °C to in excess of 2000 °C, depending upon the ceramic. During the firing process, density increases as the volume decreases, but a certain degree of residual porosity is inevitable. Reaction of the powder grains generally occurs through solid state diffusion, although silicon carbide and silicon nitride ceramics reach their final stage only through a chemical reaction process; for the nitride, the firing schedule is lengthy and conducted in a nitrogen atmosphere.

Production of the green moulding can take in several routes. The potter's clay type of operation with a damp mix is quite familiar, as possibly is slip casting, in which a slurry of powder in water is poured into a plastic mould which absorbs much of the surplus moisture, so that a damp semi-solid mass is left adhering to the mould walls. This mass is removed and fired and is typically used for making sanitary ware and vases. Other processes for handling ceramic powder mixes with liquid vehicles and binders include extrusion, injection moulding and pressing, methods comparable with those used for fabricating plastics. Flame spraying is also possible, particularly in production of silicon nitride objects.

The technology for manufacturing glass should be reasonably familiar. Components including quartz, lime and alkalies are melted in large tank furnaces and subsequently formed by drawing, casting, rolling, blowing or pressing. The well-known float glass process comprises pouring the melt onto the surface of liquid tin, on which it floats to form a uniform flat sheet of accurately controllable thickness and potentially unlimited length. Pressing and blowing operations are mostly confined to the manufacture of containers.

In any event, the glass article as first made must be annealed by heating to about 500 °C, followed by slow cooling, to relieve stresses which otherwise would make the article unreliable in service.

Joining of ceramics in low-temperature applications depends upon the use of adhesives, epoxy adhesives and 'instant' cyanoacrylates in particular being very familiar in the domestic field. Glass powders are heated to form bonds, particularly between ceramics and metals in electronic and electrical applications. Some ceramics such as aluminas can also be joined by metallising them and then brazing together what are effectively two metal surfaces.

5.7   Design factors

Designing with ceramics is dominated by the virtual absence of any ductility before catastrophic fracture occurs. This means that their low impact strength must be protected from excessive thermal shock which could generate disruptive internal stresses, and the design must be such that parts are seldom in tension. Difficulties of machining, except in the green state, are reflected in costs so machining should be avoided wherever possible by appropriate design of the component. This in turn means that the shrinkage factor during firing has to be understood and calculated so that final part dimensions are accurately predictable.

As would be expected when designing with a material which can suffer catastrophic failure on application of excessive stress, cracks, sharp edges, corners and grooves should be avoided and all section changes should be generously radiused. For the same reason a smooth surface finish should be aimed at.

6

Properties of polymers

Polymers have existed since the advent of life on earth. Natural rubber is still obtained from the rubber plant, and deposits of natural bitumen are also familiar. However, for present day engineering applications, very few natural polymers are likely to be specified, so attention must be focused on the synthetic variety, the earliest of which was a crude form of phenolic resin originating over 100 years ago. Since that time, development of polymers has posed a continual challenge to the suppliers of chemicals; the challenge also extends to designers who are perhaps not, as a group, as familiar with the ramifications of polymer technology as they are with more traditional materials. Any such difficulties as exist should be laid at the door of our educational system, despite the existence of many texts on the subject.

Polymers, as a class and in the present context, are chemical compounds based on carbon atoms; the compounds are complex, so cannot be defined as precisely as, for example, common table salt. A single polymer molecule might typically comprise many thousands of carbon atoms joined one to another, with the remaining available chemical bonds attached to other atoms such as oxygen, hydrogen, chlorine, nitrogen and sulphur, to mention just a few possibilities. To make a crude analogy, an assemblage of polymer molecules can be likened to a plate of spaghetti which has been well stirred ready for consumption. The long chain molecules of polymer are equivalent to the soft and flexible spaghetti strings, not least in their insistence on intertwining. However, unlike spaghetti, polymer molecules have side chains or areas of chemical roughness upon them, which makes it easy for the main chains to become linked together in a loose kind of knot. The unique properties of each type of polymer are attributable to the specific shape, the repeat unit structure, the distribution of molecular weights, the comparatively large sizes of the molecules and on the specific arrangement of atoms within those molecules.

As an example, we can take ethylene, normally a flammable gas, through a chemical engineering process of polymerisation under heat and pressure. The gas molecules react with each other to form enormously long chains, and transform into a waxy solid familiar to us as polyethylene plastic. The word 'poly', as a prefix to materials of this class, refers to the fact that many 'mers', that is, the individual molecular building blocks, have joined together. The term 'resin' will often be used interchangeably with 'polymer' to describe a material, this originating from a comparison between some of the early brittle polymers with natural resins. Currently, the word resin is more usually reserved to refer to solid, transparent, often brittle polymers made by polymerisation of a liquid precursor.

The large size, and consequently large surface area of individual polymer molecules, results in a high total force of attraction between groups of them; this in turn accounts for the general physical and mechanical behaviour of polymers as a class, showing the properties of reasonably modest softening points (which are never as sharp as a true melting point), a rather low hardness compared with most metals, and considerable reaction to long-term applied forces. The expression 'high' polymers is sometimes met, referring to polymers with particularly large numbers of atoms in each molecule, producing a 'high' molecular weight. As molecules become of higher molecular weight and hence longer, mechanical properties tend to improve; this, of course, is what is frequently sought by the polymer material industry to increase the utility of their products. In the context of this book, the term 'polymers' will be used in the chemical sense of the individual molecular entities, while the term 'plastics' will be reserved as a technological description for compounds or mixtures obtained when polymers are blended with other ingredients to produce a practically useful material. The word 'plastic' denotes a material condition, in which it can be softened and hardened again by sequential heating and cooling. It is also used generically to describe a member of a family, such for example as 'polyacetal plastic'. The plural is used to denote the whole universe of plastic materials, or a more restricted population representing the varieties in a single family of polymers. The relationship between polymer and plastic is the same as between flour and bread dough.

Organic polymers used in plastics and rubbers usually have few free electrons within the molecule, so they are electrical insulators, although their electrical properties may be rather complex. For the same physical reason they are generally poor thermal conductors, which in effect leads, with application of a high heat flux, to the material overheating and softening, thus imposing a limiting upper operating temperature on the material. Of course, both electrical and thermal resistances can be reduced by incorporating appropriate additives and fillers. In practice, polymers are formulated into useable plastics by incorporation of ingredients which might include fillers to occupy space and thus reduce price, reinforcing agents intended to confer additional mechanical strength, pigments for aesthetic reasons, plasticisers, stabilisers against the deteriorative effects of light, oxygen and heat, and, not least, processing aids which can increase production rates or perhaps improve the surface finish of a moulding. Despite these additives, the material is always known by the polymer designation. Thus the term 'polyethylene' standing alone may refer to a laboratory-pure specimen or to a modified version used for an engineering application. Trade names are common currency in this field; Howard (1980), among other authors quotes them by the thousand.

At present there are about 30 chemically different basic polymers used in plastics manufacture, and about 21 elastomeric polymers. Elastomers (earlier called rubbers) reflect the fundamental properties of natural rubber, which is itself a polymer. The most important and obvious property is an ability to accept high strains of perhaps several hundred per cent, and thereafter to recover totally and reasonably quickly when the load is removed. Plastics, on the other hand, are unable to accept very high strains and yet recover. There is, however, no sharp dividing line between plastics and elastomers; in fact some polymers such as polyurethanes can comprise products which can meet either description. The essence of elastomers is that their molecules are coiled or kinked, so they can physically stretch under stress, which is needed to distort the weakly attractive bonds between the polymer molecule chains and thus permit a high strain. From a molecular point of view, an elastomer is like a body-building chest expander, while a non-elastomer is like a solid, straight wire.

6.1   Types and varieties available

Plastics and elastomers are available in a number of physical forms for processing. These include coarse granules for moulding, sheets, of which natural rubber is a good example, fibres from which textiles can be woven, syrupy resins which can be chemically reacted to form hard solids, and solutions, often used as the basis for paints and adhesives. After fabrication, these materials can be converted to solid objects, thin films, foams or composite structures. These will in turn be described briefly in the following sections to familiarise the reader with some of the technical terms he will find in professional practice.

Plastics and elastomers can both be sub-divided into thermoplastic and thermosetting varieties. Thermoplastics are solids at room temperature, but can be softened by heating. Although they can, in principle, become totally liquid at a melting point like metals, in practice there exists a wide temperature range over which the material is soft and can thus be made to flow under an applied force. Subsequent cooling leads the thermoplastic to revert to the solid state without any very obvious change in the original properties. This heating and cooling cycle can be repeated and obviously provides an opportunity to shape the plastic into the desired form while in the soft state, this being the basis for fabricating articles from thermoplastics. Recovery of scrap by re-melting is an obvious possibility because the molecular chains in the thermoplastic have no particular chemical reactivity towards each other; the material melt can be held for relatively long periods, which implies that there are no appreciable material time constraints on the production process. Typical examples of thermoplastics include polystyrene, polycarbonate, polyethylene, and polyamides.

Initially, thermoset plastics appear to be like thermoplastics. They too are (often) solid materials at room temperature, and on heating they too soften. However, as softening begins, molecular chains in the polymer move relative to each other, and their inherent chemical reactivity comes into play to generate 'cross-links', which are chemical bonds between molecular chains. The result, depending upon the time and temperature profile applied, is that more or fewer cross-links will be generated to produce a lattice-like molecular structure. Although the material can be shaped during this softening stage, there is less time available for it to happen because of the speed of the cross linking reaction. After cooling, the cross-linked, yet open structure is retained and the material thereafter remains structurally solid at temperatures higher than that needed to melt it originally. So thermosetting plastics usually have higher operating service temperatures than thermoplastic materials. Unfortunately, scrap from the production process cannot be remelted as can thermoplastics, although powdered scrap can be used as a filler. Owing to the reduced time – temperature envelope available during processing, processing methods to make mouldings differ from those of thermoplastics and are reviewed in Section 6.8.

Thermosetting plastics being chemically cross-linked, result in each finished moulding being effectively a single giant molecule. On the other hand, thermoplastics may comprise amorphous structures (in which molecular chains are separate but randomly tangled through the whole material), together with discrete volumes in which crystallisation has occurred as a result of at least partial orientation of the molecular chains. The ways in which amorphous or crystalline contents affect molecular properties and processing characteristics lead to the exercise of considerable ingenuity in science and in the plastics industry, but need not worry the designer too much as long as he recognises that experts from whom he seeks opinion may invoke these factors to explain the hitherto inexplicable. Thermoset materials, of which common types include phenolics, melamine, urea, alkyds and epoxies, are continually under development to facilitate their production into ever-more complex shaped articles, to take advantage of their relatively high operating temperatures. For example, at one time, polyimide mouldings would only be supplied if manufactured by special processes operated by the polymer supplier. However, suppliers now can offer moulding compounds which can be fabricated by compression or transfer moulding in-house with little real difficulty. Even injection moulding processes are now in use, although they were first developed for thermoplastics.

6.2   Elastomers

Natural rubber was used as coatings on fabrics, to make water-proof clothing, as far back as the 1820s. In 1838 Charles Goodyear found that natural rubber could be vulcanised (cross-linked) using sulphur; the material so produced was not only highly elastic but had none of the residual tackiness characterising the natural rubber previously used. Since that time, many synthetic rubber polymers have been developed, but until recently all these needed a 'curing' or cross-linking reaction, carried out at the time of production, to achieve adequate resistance to solvents and to other environments during service. The literature abounds with descriptions of these traditional thermoset elastomers; for example Hall (1979) describes the physical and chemical structures of rubbers both natural and synthetic. Their molecular structures determine the stress – strain, viscoelastic and fatigue properties, and the effects of environments. All these factors influence satisfactory applications for each material. Hall provides a short guide to the design of rubber springs, and offers a series of examples showing how rubbers are used in rail pads, bridge bearings and vehicle suspension units.

The embryonic thermoset rubber industry soon found that a rubber-based material, whether natural or synthetic and compounded only with a curing agent, provided inadequate strength for engineering uses. Unfilled synthetic elastomers tended to be very weak, relative even to natural rubber. Various types of filler materials are now always included to improve mechanical strength and particularly abrasion resistance. Until recently, it was usual to incorporate also at least a few parts per hundred of carbon black, because this not only provided an adequate base for filling but at the same time by its darkness protected the elastomer material from the degrading effects of the ultra-violet radiation usually present in service. A recent development Anon. (1980b) concerns a synthetic filler marketed by Imperial Chemical Industries under the trade name Fortimax 423. It is based on precipitated calcium carbonate surface-coated with polybutadiene to improve adhesion to the rubber matrix. This implies that non-black synthetic rubbers might be made with the same levels of reinforcement and degree of vulcanisation as in those traditionally made with semi-reinforcing carbon black. The material is claimed to be cost effective when compared to other white filler systems which often exhibit relatively poor processing characteristics.

As happens with thermoplastics, thermoplastic elastomers are, in principle, repeatedly processable through solid, melted, solid cycles without significant deterioration. They retain their elastic properties (admittedly within a modest temperature range) without need of cross-linking reactions. Early thermoplastic rubbers were based on styrene block copolymers, as discussed by Piazza (1981), with applications for footwear soles primarily in mind. Subsequently the technology has expanded to include polyolefin, polyester, polyurethane, polysiloxane and other copolymer bases to manufacture these elastomers; some are finding particular applications in the automotive field because of increased consumer and legislative interest in safety bumpers, particularly those withstanding conventional body painting production regimes. Even nylon (polyamide) has been pressed into service as a basis for elastomers Anon. (1980*d*). This material, available in granular form, is offered in four grades of hardness ranging from 28 Shore D to 62 Shore D, and with elongation values as great as 400 %. The high impact resistance ensures that the material is usable even at −40 °C.

6.3   Cellular materials

Many, if not most, polymers can be coaxed into a porous structural form, perhaps the most familiar example being the domestic bath sponge. Some polyurethane materials are liquids, which when mixed together develop gases spontaneously to produce a foam which solidifies. The result is a rigid or flexible porous structure. In other cases, solid polymers can be mixed with 'blowing agents' which will generate gas at a predetermined temperature. Subsequent heating of the mixture to the softening temperature of the plastic then enables this gas to be produced and held by the melt; the degree of expansion of the foam is usually controlled by containing the mixture in a mould with predetermined dimensions. Cellular materials may be of open or closed pore structure, depending whether it is necessary to absorb water, as in the case of a sponge, or whether it is essential to contain still, entrapped air, as in the case of insulating boards. In almost every case foams have a solid skin which forms where the material touches its containing mould during manufacture. If open pores are needed at the surface then the foam bun is sliced mechanically, as happens with polyurethane foams used for making cushions and mattresses.

Perhaps the most familiar use of rigid cellular plastics is in the building industry, where expanded polystyrene or polyvinylchloride are used in board form for insulating roof,spaces, and foamed urea formaldehyde or polyurethane granules are used for filling cavities between walls. Foamed polyurethane is produced as a sandwich between layers of thick paper for insulation boards where a finish appearance is required. Details of such applications have been given by the Building Research Establishment, Anon.(1979)

6.4 Plastic film

Plastic film is vital to the domestic scene, being used as a moisture barrier for wrapping food and as plastic bags. Recently, rather thick film has appeared as blow-moulded bottles for soft drinks. The packaging industry has a very clear picture of the optimum use areas for plastic films for wrapping, and polyethylenes of various kinds, polypropylene, cellulosics and polyvinylidene difluoride are used for protecting a very wide range of consumer goods. Sometimes a pure polymer film is used, at others (as in the case of some soft drinks) a laminated construction is necessary to limit gas diffusion in or out. Films are often required to excel as moisture barriers, and may be expected to act as oxygen barriers, with, perhaps, exclusion of light as a further function. Films can be produced by extrusion, most often in the form of a thin-walled tube: internal air pressure expands the tube which is slit and rolled up as film.

6.5 Costs relative to other materials

It has already been stated in the discussion on metals that the cost of a material purchased ready for processing is only one item in the total cost of the final product. That item may or may not be significant, depending upon the complexity of the production process and upon the associated indirect costs of the organisation doing the work. Historically, conventional engineering materials including steel, yellow brass, zinc, aluminium, magnesium and engineering plastics have had chequered but ever-more important careers since the early part of the twentieth century. Although there were occasions when the ranking order of costs changed for these materials, in general, an ever-rising price increase year by year tends to keep a consistent pecking order. Thus in 1982, of the materials just listed, yellow brass was the most expensive per unit volume and hot rolled sheet steel was the cheapest, with engineering plastics only very little more expensive than steel. It can be predicted that, until the end of this century, the comparison offered here will remain valid, and that engineering plastics will continue to perform very well from a cost point of view compared with metals calculated on a cost-per-unit volume basis. This kind of market intelligence is, of course, vital to a design engineer considering alternative part geometries to achieve a given strength requirement. From this point of view, engineering plastics will very often outperform metals, primarily owing to their high stiffness coupled with relatively low density. The variable market situation on prices of plastics precludes offering much detail of costs even at a spot point in time, although Table 6.1 provides some information as of the end of August 1984. Perhaps less changeable are relative costs of plastics, although here also values can be influenced by the efforts of materials suppliers to reduce the price of their formulated plastics by increasing use of ever cheaper additives, even though the base polymer itself may increase in price. Specimen information is offered in Table 6.2.

From time to time information relevant to the costs of plastics and their processing appears in the technical press. West (1984) discusses a costing procedure for polymer processing, and Walshe & Lowe (1984) report a computer-aided cost estimation method for the benefit of tool makers producing injection moulds.

Table 6.1. *Prices of some plastics — Germany 1984*

Material	Market price range (US$)
Polyethylene, high density	0.66 – 0.70
Polyethylene, linear low density	0.73 – 0.77
Polyethylene, low density	0.68 – 0.70
Polypropylene	0.75 – 0.79
Polystyrene	0.84 – 0.87
Polyvinyl chloride	0.56 – 0.58

Table 6.2. *Relative costs of formulated plastic moulding powders (1984)*

Material	Relative cost by volume	Specific gravity	Relative cost by weight
Polyethylene	1.0	0.9	1.0
Polypropylene	1.0	0.9	1.0
Polyvinylchloride	1.5	1.4	1.0
Polymethylmethacrylate	2.0	1.2	2.0
Acrylonitrile – butadiene – styrene	2.5	1.0	2.5
Polyethylene isophthalate	2.5	1.1	2.5
Polyester (bisphenol)	3.0	1.1	3.0
Furane	3.0	1.2	3.0
Polyamide (Nylon 6 and 66)	3.5	1.1	3.5
Polyphenylene oxide (modified)	3.5	1.1	3.0
Polyacetal	4.0	1.4	3.5
Polycarbonate	4.0	1.2	3.5
Epoxy	4.0	1.1	4.0
Polyamide (Nylon 11)	7.0	1.0	7.0
Polysulphone	12.0	1.2	11.0
Poly(phenylene sulphide)	13.0	1.3	10.0
Polyvinylidene difluoride	23.0	1.8	13.0
Poly(tetrafluoroethylene)	26.0	2.2	12.0

6.6 General properties of plastic materials

Inclusion of plastics and elastomers as materials of construction can considerably widen the designer's horizons, as these materials offer a widely spread range of parameters, and indeed in some can surpass the performance of more traditional materials. For example, density values as low as 16 kg/m^3 can be reached in highly expanded foams; addition of heavy fillers can yield a density as high as 2500 kg/m^2. Modulus values can range from 1 MN/m^2, to 200 GN/m^2 for a carbon-fibre reinforced composite. Permissible strain in use may range from 1 % in some reinforced plastics to 100 % for a natural rubber.

Furthermore, there are some qualitative differences in behaviour between polymeric materials and other engineering materials. Much higher strains may be accepted. Even for load-bearing plastics, strains may vary from $1-4$ %, although such levels might lead to non-linear load/deflection behaviour through consequent changes in geometry. This response can be allowed for by the designer, particularly as the limiting design criterion might often be distortion rather than maximum available strength. Again, because of the mode of lay-up of composites, or owing to the peculiarities of polymer melt flow into an injection mould, there may be no isotropy of properties in a manufactured component. This is perhaps best seen in the case of a carbon-fibre-reinforced plastic, which in the same sample might have E_1 of 200 GN/m^2 in one direction, and E_2 of 50 MN/m^2 in a different direction. Indeed, a whole field of structural analysis has been developed to take advantage of these characteristics. Even in the case of a simple, unfilled injection-moulded part, the act of forcing a softened, long molecular chain polymer through a small orifice into an open mould often results in some orientation of those chains so that in the flow direction tensile properties will exceed those in a direction at right angles. When one considers that most injection moulded objects are of complex shape, with the flow pattern being correspondingly complex, the problem of calculating what the critical strength might be in a given cross-section is of mind-boggling difficulty. It may readily be understood that a whole science of measuring and displaying stresses in moulded and machined objects has developed.

The following tables provide representative properties of some of the main classes of thermoplastics, thermosetting plastics, thermoplastic elastomers and foamed plastics met with in engineering practice. Values quoted must, of course, be appoximate because of the influence on them of many factors. These include the molecular weight of the test sample, the extent of incorporation of additives and fillers of various kinds, small chemical and physical changes made to the polymer for competitive reasons by the supplier, a variety of different test methods to express ostensibly the same physical parameter, variations in test results due to the effects of strain rate, ambient temperature, humidity and flow pattern in the production of each test piece and many more factors besides. For these reasons, only generalised values are quoted, sometimes over a range, in the expectation that they will be used only for cursory comparisons. Accurate figures are available from suppliers and must of course be used for serious design work. Indeed, some suppliers are very helpful in this respect and offer complete design data books for specific materials such as a brand name polyamides or polyacetals, containing specimen calculations and the particular design rules which must be adopted to ensure satisfactory performance in the product.

Table 6.3 illustrates the properties of some engineering thermoplastics, referring only to unfilled and unreinforced base polymers. Higher values of some of the properties would be expected if additives were incorporated.

Table 6.4 offers data on properties of fully formulated thermosetting plastics; these have included in them various reinforcing agents which alone ensure suitability for serious engineering use. Without them the materials would suffer excessive shrinkage on cooling from the moulding process, causing severe internal strains and possibly internal cracking.

Certain aspects of thermosplastic and thermosetting polymer performance should be emphasised because of their importance to the designer. It should be recognised, for example, that the phenomena of stress relaxation are directly dependent upon time, temperature, applied stress and, in the case of a few materials, ambient or internal humidity. As an example, it can be shown that under simple flexure, the deflection of a polypropylene component may increase by 120% if the applied load is doubled, or, alternatively, may increase by 80% if the temperature is increased from 20 °C to 60 °C, but the deflection may increase by as much as 400 per cent if the applied load remains in place at ambient temperatures for a year. Furthermore, all these materials become increasingly brittle at low temperatures when partial crystallisation occurs. Brittleness is also a consequence of surface degradation through oxidation at high temperatures or through the effect of ultra-violet light during normal weathering. Oxidation effects are accelerated by increasing temperature, which itself may modify other properties. It is well to recall the low specific heat and low thermal conductivity of most unfilled plastics because under light or heat irradiation, surface temperatures can rise very rapidly, with a consequent risk of deterioration. On the same theme, it should be remembered that thermal expansion effects are an order of magnitude higher than for metals; in composite structures it may be essential to allow for this to permit differential movement. Particularly where there are mechanical fastenings and joints, it is possible that thermally induced movements could set up unacceptable stress concentrations.

Table 6.3. *Properties of some engineering thermoplastics*

Material	Elongation at break (%)	Flexural strength (N/mm²)	Notched impact (J/cm²)	Hardness (Rockwell unless marked)	Thermal conductivity (W/m K)	Heat deflection temperature (°C, 4.6 kg/cm² applied load)	Coefficient of thermal expansion (× 10⁻⁶/K)
Acetal (copolymer)	20–75	118	0.38	R118	1.32	158	130
Acetal (homopolymer)	40	97	0.30	R120	1.32	170	90
Polyamide (nylon 6)	20–360	34	0.42	R119	1.39	170	80–100
Polyamide (nylon 6/6)	40–80	28	0.32	R108–118	1.39	190–240	90
Polyamide (nylon 11)	300–350	10	0.38	R108	2.30	145–155	90
Acrylonitrile–butadiene–styrene	10–15	49–88	1.46	R85–110	1.51	95–107	60–110
Polycarbonate	80–120	93–245	3.11	R120	1.10	140–145	60–70
Polyethersulphone	50–100	126	0.34	M88	0.91	240	55
Polyethylene terephthalate	4	80	0.18	M85	1.34	105	70
Polyphenylene oxide (modified with styrene)	60	93	1.04	R117	1.25	150	60
Polyphenylene sulphide	1.6	96	0.06	R124	1.63	135	49
Polypropylene	200–700	39–59	2.23	R80–110	6.72	125	110
Polysulphone	50–100	98–118	0.25	R120	6.72	185	30–55
Polytetrafluoroethylene	250–400	14	0.62	Shore D 50–65	1.44	121	55
Polyvinyl chloride (unplasticised)	60	78	2.08	Shore D80	1.03	70	67–90

Table 6.4. *Properties of some thermosetting plastics*

Material	Flexural strength (N/mm^2)	Izod impact strength (J/cm^2)	Hardness (Rockwell unless marked)	Thermal conductivity (W/m K)	Maximum continuous working temperature (°C)	Coefficient of thermal expansion $(\times 10^{-6}/K)$
Alkyd	70 – 140	1.73	70 – 75 (Barcol)	4.8	149	20 – 50
Diallyl phthalate	80 – 180	1.62	E60 – 65	3.8	200	10 – 42
Epoxy	40 – 470	3.35	M100 – 110	3.6	105 – 160	20 – 45
Polyester dough moulding compound	70 – 280	3.69	50 – 80 (Barcol)	3.1	120 – 205	20 – 50
Melamine – formaldehyde (glass – filled)	110 – 160	1.96	M110 – 125	3.6	150	20 – 50
Phenol – formaldehyde	70 – 400	1.92	R150	2.6	205	19 – 25
Urea – formaldehyde	70 – 130	0.69	M115 – 120	2.0	75	20 – 60
Polyimide	69 – 248	11.58	M118	2.2	250	10 – 27

Plastic mouldings which are homogeneous to the naked eye, may in fact not be so because of the orienting effect of the material flow through the mould gate during filling. Such effects are of increased importance if the material is reinforced with fibres, since not only can the fibres become preferentially oriented within the work piece, but they can be broken, with the length reduced by as much as one order of magnitude, during the injection process. The result is that the performance of the piece in service may be far less than expected from test results on more carefully prepared, and perhaps differently prepared, laboratory test specimens.

Like all organic (carbon-based) substances, polymeric materials are combustible, although the extent of this as a problem depends upon the individual material and on any additives incorporated in it. This aspect of materials performance is sufficiently important to the designer to warrant wider discussion; the subject is accordingly dealt with in Section 8.5.

In contradistinction to metals and ceramics, the chemical resistance offered by plastics is different and in general, complimentary. Thus most plastics are minimally affected by mineral acids and alkalies, although some are more sensitive than are metals and ceramics to the effect of organic acids and organic solvents.

Thermoplastic elastomers offer the ease of fabrication of thermoplastics combined with the flexibility and toughness expected from the more conventional cross-linked or vulcanisable elastomers. Generally, the thermoplastic elastomers as a class bridge the gap very neatly between the hardness of conventional elastomers (typically Shore Durometer A values of 20−95), and that of the much harder thermoplastics (with Shore Durometer D values of 60 upwards). Thermoplastic elastomers, like rigid thermoplastics, may be used in some cases with no additives, although extenders, plasticisers and solid fillers are often incorporated for technical or commercial reasons. Although thermoplastic elastomers can be used as a class between temperature limits of −70 °C to +150 °C, nevertheless they should never be considered for use at temperatures approaching that of processing. Table 6.5 shows some comparative properties of a number of elastomers which are identified specifically as vulcanisable or thermoplastic. In all cases, these figures apply to elastomers compounded with fillers and other additives to the state in which they would normally be presented to a customer. As with all tables of properties in this book, the values are meant to be illustrative only, and the designer is advised to ask materials suppliers for specific data once he thinks a reasonable materials choice has been made.

Some of the caveats already mentioned may leave the designer with a rather poor idea of the potential of plastics in engineering usage. Their deficiencies are likely to be most obvious at the limits of design, where there may have been a

Table 6.5. *Comparative properties of elastomers*

Type of polymer V = Vulcanisable T = Thermoplastic	Material	Specific gravity	Rockwell hardness	100% tensile (N/mm^2)	Ultimate tensile (N/mm^2)	Elongation at break (%)	Tear strength (Kg/cm)	Compression permanent set (%)	Brittleness temperature (°C)
V	Butadiene rubber	1.04	65	2.5	15.0	550	18	65	−73
V	Isoprene rubber	1.04	57	3.0	34.0	750	83	20	−58
V	Styrene–butadiene rubber	1.07	67	3.5	21.5	730	35	30	−47
V	Acrylonitrile–butadiene rubber	1.08	72	5.5	22.0	700	34	55	−58
V	Natural rubber	1.06	65	1.8	27.0	850	60	60	−58
V	Acrylic rubber	1.25	68	5.3	13.0	500	28	88	−18
V	Chlorinated polyethylene	1.28	73	4.0	18.0	620	62	80	
V	Chlorosulphonated polyethylene	1.26	72	6.8	25.0	480	65	68	−45
V	Chloroprene rubber	1.25	72	6.5	24.0	950	65	30	−37
T,V	Urethane rubber	1.26	70	7.0	30.0	525	80	40	−43
V	Epichlorhydrin rubber	1.42	75	7.3	20.0	900	54	50	−35
V	Polysulphide rubber	1.47	67		8.5	350		90	−45
V	Fluorinated rubber	1.77	73	5.0	18.0	350	27	3	
T,V	Ethylene propylene rubber	1.30	70	4.0	19.0	650	50	4	
V	Silicone rubber	1.17	65		7.5	350	17	10	−62
T	Propylene oxide rubber		47	4.2	24.0	825	30	70	−65

major effort to save weight or cost. However, provided that applied strains are sufficiently low that long-term creep will not constitute a problem, and provided that the limiting upper service temperature is allowed for, plastics will provide many years of service, dependent upon their rate of degradation in the ambient environment. Even creep itself can be beneficial, as it provides relief of high local stresses which otherwise might cause cracking, so that, for example, bearing pads for bridges will conform accurately to the rigid surfaces with which they are in contact. Furthermore, interference fits are more easily accomplished provided that the material has a combination of low modulus and high thermal expansion. Controlled anisotropy of properties can be achieved in certain mouldings and can lead to more efficient designs. Lastly, most plastics offer low coefficients of friction against metals and each other, and under low-load and low-speed conditions can be used as effective sliding or bearing surfaces subject to very little wear under poor lubrication conditions.

At some stage of his education, the designer would do well to accept that much of the jargon of plastics centres around the use of trade names. Organisations such as RAPRA Technology Ltd of Shrewsbury, England, publish extensive collated lists of these, relating trade name to the basic polymer and to the supplier using it. Such lists are of very great utility, although their use should not blind the designer to the fact that, within a given family of materials such as polyamide 6/6, the envelopes of properties may not be congruent. Different trade names for, ostensibly, the same family of materials are not simply a marketing ploy, but are a very useful aid to communication in the designing, engineering and materials professions.

6.7   Some specific aspects of plastics

This is not the place to discuss in any detail all the manifold properties of plastics which make them appealing for design use. More than enough will be found in text books on plastics and particularly in literature from plastics supply houses. However, a few key properties are worth highlighting with a brief commentary.

6.7.1   *Stiffness and toughness*

Stiffness, or flexural modulus as it is more commonly called, is an important attribute of most materials under consideration for engineering applications. Metals in general have a flexural modulus range of the order $10-70$ GPa. Over the years, a steady progression of improved engineering plastics has taken the design world from polyamide in the late 1940s via polycarbonate and polyacetal to glass-reinforced materials such as the polyesters and polyphenylene oxide of the 1980s. All these had successively improved stiffness, so that by perhaps 1990, some plastics yet to be developed could be as stiff as those metals in the lower flexural modulus range, Pastorini (1984).

ICI (1984) have demonstrated clearly the relation between stiffness and toughness of engineering plastics, as outlined in Figure 6.1. This confirms that the basic properties of some polymers are essentially rather poor, but addition of mineral fillers or rubber modifiers considerably increases impact strength and particularly the flexural modulus; glass fibre reinforcement, together with rubber modification, improves the picture still more. The diagram incorporates the properties of polypropylene, polymethyl methacrylate, polyacetal, polycarbonate, polyethyl ether ketone, polyethersulphone and polyamide 6/6.

6.7.2   *Ultimate tensile strength*

This is a material property which seems ubiquitous. It is quoted everywhere in tables of properties, probably through being simple to quantify

Fig. 6.1. Stiffness/toughness relationship of engineering plastics.
(Copyright: 1985 Standard Telephones and Cables plc.)

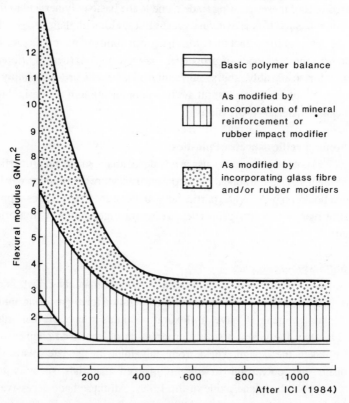

After ICI (1984)

Notched Izod impact strength at 23°C (J/m)

from easily made test pieces. However, the designer of engineered structures, while taking cognisance of ultimate tensile strength, would not expect normally to apply stresses anywhere near this limit. It is more usual for structures to work within the Hookean (elastic) limits of any material. For plastics, this may be limited to a very few per cent of strain, above which significant creep would begin and continue to occur over periods measured up to years, depending upon the temperature envelope in which the material operates. It is for this reason that the designer should not be misled by an apparently high ultimate tensile strength claimed for a plastic, since much of its potential is likely to be unavailable to him unless the product has a very short service-life expectancy.

Some representative tensile strength values are offered in Figure 6.2, showing that many of the engineering plastics find themselves towards the lower end of the spectrum. The misleading nature of ultimate tensile strength as a design parameter can be seen from the fact that porcelain and stoneware have a lower tensile strength than most of the possible competitive plastics, and yet these ceramics have been used perfectly satisfactorily through millennia for applications into which plastics only now are beginning to find their way.

So, summarising, by all means take note of ultimate tensile strength but with plastics do not be tempted to design anywhere near to their limits except for

Fig. 6.2. Tensile strength values of engineering materials. (Copyright: 1985 Standard Telephones and Cables plc.)

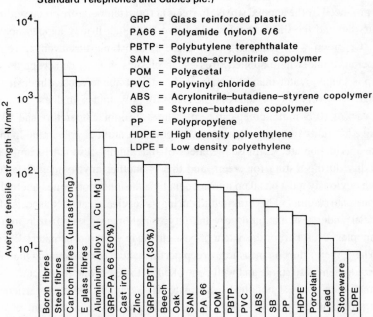

GRP	= Glass reinforced plastic
PA66	= Polyamide (nylon) 6/6
PBTP	= Polybutylene terephthalate
SAN	= Styrene–acrylonitrile copolymer
POM	= Polyacetal
PVC	= Polyvinyl chloride
ABS	= Acrylonitrile–butadiene–styrene copolymer
SB	= Styrene–butadiene copolymer
PP	= Polypropylene
HDPE	= High density polyethylene
LDPE	= Low density polyethylene

special short-term purposes; always apply a conservative derating factor. It should be remembered that tensile test methods themselves vary; a test applied at a suitable strain rate for a metal may yield an unrepresentative result when applied to a plastic, owing to the effects of ambient temperature and creep on the latter. Furthermore, materials of high ductility will generate a considerably reduced cross-sectional area at the point of break, and this area is not only difficult to measure exactly, but is of doubtful validity as a means of calculating an ultimate tensile strength expressed in terms of force per unit area.

6.7.3   Creep

The topic of creep is too extensive to be discussed even in broad detail in the present book, but its existence must not go unnoticed. If a material is to be exposed to a service condition involving a combined stress and temperature envelope with significant random variations, or cyclically throughout the product life, then it should be expected that creep and consequent dimensional changes will occur in organic materials; creep has to be considered even when designing metallic objects such as parts in a steam power plant. In an extensometer type tensile test (in which the two ends of a test bar are pulled apart by a measured force to a measured extent), it can commonly be seen that after an increment of load has been applied, the test piece continues to extend for an appreciable time. This time may be so long that the tensile test cannot be performed in the way planned, as there is no definite end point for the elongation produced by the applied stress. In steels at room temperature creep seldom occurs beyond the 0.2% proof stress value, although it may happen above 200 °C; creep will, however, take place in most plastics even at room temperature under high applied stresses. In plastics, it must be accepted that creep is much greater than in metals, and can continue to cause dimensional changes over very long periods of time. Thus for polyamides, under stresses of the order of 10 N/mm^2, creep can be measured at room temperature and 50% relative humidity over periods up to 10^5 hours. Mention of polyamide in this context is deliberate, as most members of that family respond to ambient humidity during testing for creep, and this is another environmental factor which obviously must be taken into account by a designer concerned about the dimensional stability of his brain-child during its service life. Other aspects of a material under test, including crystallinity, moulding stresses frozen into a thermoplastic test bar, or the extent of post-curing in a thermoset material, will each have their effect on measured creep performance, even assuming that the test equipment is so stable and well controlled as to be beyond reproach.

Creep data are best obtained from materials suppliers for their particular grades of plastics, but the designer must be cautioned to be certain of the relevance of the test method used to obtain those figures on which he is going to rely for his design. Brown (1981) provides a useful text on current creep test methods.

6.7.4   *Environmental effects*

This subject in its relation to plastics is one which, in the literature, tends to be either much too generalised or, very often, much too specific. Considerable information is presented in text books in the form of lists of generic types of plastics and their weather resistance, often expressed by a brief phrase such as 'slight-to-moderate deterioration', or 'fair-to-good when stabilised'. More specific information, on the other hand, may relate to a particular commercial grade of a material, and provide figures giving the effect of ageing in an artificial weathering machine against a given end point. Resistance of many plastics to organic solvents is also often described in generalities such as 'generally good resistance' or 'attacked by (specified) solvent'. Numerical data are only available when a particular solvent initiates stress cracking in a sample exposed to a suitable test regime, and here the time to stress cracking can be defined in hours or days. Unfortunately the figure may be of little use to a designer in considering the likely performance of his new product unless it spends its life in that solvent.

Chemical resistance of plastics is also widely expressed in subjective terms, in which resistance to dilute or concentrated acids, alkalies, oils or greases, may be specified as 'poor', 'satisfactory' or 'good'. Again this is too indefinite for somebody with a particular application in mind for a special commercial grade of a single plastic type. Here, as always, recourse must be had to the material supplier, although it is possible, with the exception of grades of plastic to be used for food purposes, that the information specifically required may not exist and might need to be specially obtained for the designer, with consequent cost and time penalties.

So we have seen that plastics can react to weathering or to a chemical environment, but results observed on representative families cannot be translated into very significant design data. There is no substitute for special tests made under the likely working circumstances for the candidate materials. On the other hand, he who selects plastics can always play safe by using types known to offer satisfactory performance in other industries, although such a choice may not necessarily be the best for his purposes. For example, in a projected alkaline service environment one might look to the established choice of polytetrafluoroethylene, which offers good resistance against either dilute or concentrated alkalies. However, in a different application, the remaining properties of PTFE within its materials envelope may not be important, and indeed may prove to be expensive luxuries; it could happen that polyethylene would meet the function just as well, and at considerably reduced price.

Table 6.6. *Common fabrication methods for thermoplastics*

Manufacturing methods / Thermoplastics	Blow moulding	Calendering	Casting	Compression moulding	Extrusion	Injection moulding	Laminating
Acrylonitrile–butadiene–styrene		X			X	X	X
Acetal resins	X				X		
Acrylic resins	X		X	X	X	X	X
Cellulose acetate	X		X	X	X	X	
Cellulose acetate butyrate	X		X	X	X	X	
Cellulose propionate	X		X	X	X	X	
Chlorinated polyester					X	X	
Ethylene copolymers	X				X	X	
Ethyl cellulose				X	X	X	
Fluoro plastics					X	X	
Methyl pentene	X				X	X	
Nylons	X		X		X	X	
Polyamide–imide				X		X	
Polyaryl ether					X	X	
Polyaryl sulphone					X	X	
Polybutylene	X				X	X	
Polycarbonate	X				X	X	
Polycarbonate–ABS					X	X	
Polyethylenes	X				X	X	
Ionomer	X				X	X	
Phenylene oxide based materials					X	X	
Polyphenylene sulphide						X	
Polyimide				X			
Polypropylenes	X				X	X	
Propylene copolymers	X				X	X	
Polystyrenes	X				X	X	
Polysulphone					X	X	
Polyvinyl chloride	X	X			X	X	
Polyvinyl chloride copolymers	X				X	X	
PVC–acrylic compounds					X	X	
PVC–ABS compounds		X			X	X	
Styrene–acrylonitrile					X	X	
Thermoplastic polyesters	X				X	X	

Rotational moulding	Slush moulding	Stamping	Vacuum forming	Pressure forming	Mechanical forming	Cold forming	Sintering	Thermoplastics
								Acrylonitrile – butadiene styrene
X		X	X	X	X	X		
X								Acetal resins
X			X	X	X			Acrylic resins
			X	X	X			Cellulose acetate
								Cellulose acetate
X			X	X	X			butyrate
			X	X	X			Cellulose propionate
								Chorinated polyester
X			X	X	X			Ethyene copolymers
			X	X	X			Ethyl cellulose
			X	X	X	X		Fluoro plastics
			X	X	X			Methyl pentene
X			X	X	X	X		Nylons
							X	Polyamide – imide
								Polyaryl ether
								Polyaryl sulphone
X			X	X	X			Polybutylene
X		X	X	X	X	X		Polycarbonate
X			X	X	X			Polycarbonate – ABS
X			X	X	X			Polyethylenes
X			X	X	X			Ionomer
								Phenylene oxide
		X	X	X	X			based materials
								Polyphenylene
								sulphide
							X	Polyimide
		X	X	X	X			Polypropylenes
			X	X	X			Propylene copolymers
X			X	X	X			Polystyrenes
			X	X	X	X		Polysulphone
						X		Polyvinyl chloride
								Polyvinyl chloride
X	X		X	X	X			copolymers
								PVC – Acrylic
			X	X	X			compounds
			X	X	X			PVC – ABS compounds
			X	X	X			Styrene – acrylonitrile
								Thermoplastic
			X	X	X			polyester

Table 6.7. *Common fabrication methods for thermosets*

Manufacturing methods Thermosetting	Casting	Centri-fugal moulding	Compression moulding	Filament winding	Injection moulding	Laminating	Matched die moulding
Alkyds			X		X		
Allyl diglycol carbonate	X						
Diallyl phthalate	X		X		X		
Epoxy resins	X	X	X	X	X	X	X
Melamines			X				X
Phenolics	X		X		X	X	
Polybutadienes	X		X			X	
Unsaturated polyesters	X	X	X	X	X	X	X
Silicones	X		X		X	X	
Ureas			X		X		
Polyurethanes	X		X		X	X	
Polyimides							

6.7.5 Transparency

The attainment of transparency is often a problem for designers. Acrylics, of course, have been well known since the days of the Second World War, and from that time several polymers have been developed, with the prospect of at least a degree of transparency, albeit with the penalty of a certain amount of haze or of inherent colouration. Selection of such a plastic depends on balancing the price which can be accepted to achieve optical clarity, compared with the lower price possible, with less adequate technical performance, if a slightly hazy or coloured material were chosen. Such a material might function for a liquid sight glass. Whelan (1981) discusses 16 different types of transparent plastic, providing information on the class of material, a range of trade names and suppliers, a 1981 price indication and data on and other factors such as percentage transparency, percentage haze and mechanical and thermal properties. An important section of Whelan's paper deals with moulding qualities and colour potential. The materials described include acrylics, cellulose acetate, cellulose acetate−butyrate, cellulose propionate, aromatic polyamide, polyarylate, polycarbonate, polyethyleneterephthalate, poly-methylpentene, polystyrene, polysulphones, unplasticised polyvinylchloride, styrene−polyacrylonitrile, styrene−butadiene−styrene block copolymer and nylon 6. Some of these materials are completely colourless and others have tints deepening down to amber.

Pultrusion	Transfer moulding	Wet lay-up moulding	Reaction injection moulding	Reinforced reaction injection moulding	Polymerisable monomer reactant resins (PMR)	Thermosetting
	X					Alkyds
						Allyl diglycol carbonate
	X					Diallyl phthalate
X	X	X				Epoxy resins
						Melamines
	X					Phenolics
	X	X				Polybutadienes
X	X	X				Unsaturated polyesters
	X					Silicones
						Ureas
			X	X	X	Polyurethanes
					X	Polyimides

Pye (1982) points out that an advantage transparent plastics have over common glass is the combination of impact strength, toughness and lightness. A major problem, even today, with plastic glazing for vehicles is the problem of surface scratching from high-velocity dust. Although this problem has been solved adequately by the application of silicone-based surface films, there is still scope for improvement in resistance to abrasion, even though these films are spectacularly effective. It should be realised that not all transparent plastics are necessarily thermoplastic. For example, allyl diglycol carbonate combines the optical properties of glass with typical thermoset resin properties; its resistance to welding sparks and to impact is exploited in prescription eyeware and safety glasses. Furthermore, radiation-cured silicone rubbers can be totally transparent as well as elastomeric. Compounds of a similar feel and appearance, which absorb moisture and then become slightly flexible, are used in modern contact lens technology.

6.8 Fabrication routes

The variety of fabrication routes for thermoplastics and thermosetting plastics is now bewildering if one also considers the variants on the main processes, which are themselves numerous enough as shown in Tables 6.6 and 6.7. Broadly speaking, manufacturing methods for thermoplastics involve taking a preformed solid object and heating it so that a new shape can be impressed upon it, or (more commonly) granules of the plastic are softened by heat, and the melt is injected into a mould, or extruded to form the required section; air pressure may be used to blow an extruded tube into a thin film. In the case of thermosetting plastics, only the more sophisticated injection technologies can run to the luxury of handling a melt of the material because of the fast reaction times which lead to curing via the cross-linking processes. A certain amount of reaction takes place in, for example, transfer or compression moulding. Another significant manufacturing process, for composites, is to wind reinforcing filaments onto a mandrel together with liquid resin, or to use a flat-surface variant. The modern reaction injection techniques are noteworthy because the plastic in question does not pre-exist before the moulding operation. Instead, two or more liquid precursors of the material are very rapidly mixed at the point of injection into a mould, and a totally new material then emerges as a moulded entity after the curing cycle is complete.

Even a modest discussion of plastics forming methods, which a designer may care to exploit, could (and often does) occupy a whole book of itself. For our purposes, Table 6.8 illustrates sufficient salient features of the more important techniques for a designer to arrive at some idea of the production route most suitable for his new product.

Sometimes conventional methods of fabricating plastics are inadequate, and it becomes necessary to machine all or at least a significant part, of a plastic article. Machining Data Handbook (1980) provides quantitative data on turning, drilling, milling and band-sawing for a limited range of materials. Such information is very sparse indeed compared with that available for metals. Howard (1980) provides a general table of machining qualities for a wide range of thermoplastic and thermoset materials, pointing out that plastics inherently have different properties from wood and metals, initially causing difficulties in machining if using the same tool settings as for traditional materials. The chief of these is that the coefficient of thermal expansion of plastics is roughly an order of magnitude greater than for metals, so that binding of tools and cutting blades can occur. Thermal conductivity is substantially less than for metals, so machining speeds must be relatively low to avoid surface melting; furthermore the elastic modulus is much smaller than that of a metal, and elastic recovery is noticeably large. Consequently, because of these problems of heat dissipation and dimensional changes, of tool clearances, tool material, tool angles and machine speeds, considerable experimentation and adaptation is often needed to attain reasonably machined surfaces. Nevertheless, it is possible to machine plastics, but the designer should not depend upon the existence of a well-established technology able to meet his possibly casual specification 'machined finish'.

Table 6.8. *Features of common fabrication processes for plastics*

Process	Advantages	Limitations
Blow moulding	Low tool and die costs High production rates Low scrap rates Produces one-piece complex hollow shapes	Restricted to hollow articles Essentially a two step process (hollow tube necessary) Wall thickness difficult to control
Casting	Large articles can be moulded Low mould costs Low finishing costs	Labour cost high Restricted to simple shapes
Compression moulding	Large component can be produced Little material waste	Tooling costs can be high Close tolerances difficult to achieve
Extrusion	Low testing costs High production rates Variety of complex cross sections can be produced	Close tolerances are difficult to achieve Parts must have uniform, even section
Injection moulding	High production rates Parts have good surface finish Complex shapes can be produced	High tool and die costs Large production runs necessary for economy Size of parts limited
Rotational moulding	Low tooling costs Large area parts can be produced No seams	Slow production rates Cumbersome material loading and part removal
Vacuum forming of sheet materials	Low cost technique Thin cross section parts can be produced	Restricted to articles with low profile
Pressure forming of sheet materials	Used for deep drawn articles Can be used on sheet material too thick for vacuum forming. High production rates	Expensive process Highly polished moulds required
Mechanical forming of sheet materials	High production rates	Restricted to simple shapes

Table 6.8. *Features of common fabrication processes for plastics (continued).*

Process	Advantages	Limitations
Cold forming	High production rates Low cooling costs	Thin parts not economical
Transfer moulding	Rapid production rates Thin sections are possible Delicate inserts may be moulded in-situ	Moulds complex and expensive Component size limited
Slush moulding	Low mould costs Can produce fairly intricate articles with good finish	Slow production rates

Table 6.9. Assembly methods for thermoplastics

Assembly methods Thermoplastics	Adhesive bonding	Dielectric welding	Induction bonding	Mechanical fastening	Solvent welding	Spin welding	Staking	Thermal welding	Ultrasonic welding
Acrylonitrile butadiene styrene	X			X	X	X	X	X	X
Acetal resins	X		X	X	X	X	X	X	X
Acrylic resins	X			X	X	X			X
Cellulose acetate	X				X	X	X		
Cellulose acetate butyrate	X				X	X	X		
Cellulose propionate	X				X	X	X		
Chlorinated polyether	X			X				X	
Ethylene copolymers		X						X	
Fluoro plastics	X								
Methyl pentene			X				X		X
Nylons	X			X					X
Polyamide – imide	X			X					
Polyaryl ether	X			X			X		X
Polyaryl sulphone	X			X					X
Polybutylene							X		X
Polycarbonate	X			X	X	X	X	X	X
Polycarbonate – ABS	X			X	X	X	X	X	X
Polyethylenes	X		X	X		X	X	X	X
Ionomer		X						X	
Phenylene oxide based materials	X			X	X	X	X	X	X
Polyphenylene sulphide	X			X					X

Table 6.9. *Assembly methods for thermoplastics*

Assembly methods / Thermoplastics	Adhesive bonding	Dielectric welding	Induction bonding	Mechanical fastening	Solvent welding	Spin welding	Staking	Thermal welding	Ultrasonic welding
Polyimide	X			X					
Polypropylenes	X		X	X		X	X	X	X
Propylene copolymers	X		X	X		X		X	X
Polystyrenes	X		X	X		X	X	X	X
Polysulphone	X			X	X		X		
Polyvinyl chloride	X	X	X	X				X	
Polyvinyl chloride copolymers	X	X	X	X					
PVC – acrylic compounds	X	X	X					X	X
PVC – ABS compounds	X							X	
Styrene acrylo nitrile	X		X	X	X	X	X	X	X
Thermoplastic polyesters	X			X	X	X	X	X	X

After forming the various parts of a product, there is of course the need for final assembly. Tables 6.9 and 6.10, respectively, illustrate some common assembly methods for both thermoplastics and thermosetting plastics. Many of the methods for thermoplastics depend upon the solubility of these materials, as in the cases of adhesive bonding and solvent welding, or, alternatively, depend upon the possibility of surface melting to cover the various aspects of welding. Assembly of thermosetting plastics offers fewer options due to their thermally intractable and insoluble nature. In general, adhesive bonding of suitably prepared thermosetting surfaces is feasible, and mechanical fastening is always a viable option.

A major expense of the injection moulding process to produce thermoplastic articles is the manufacture of the mould itself, with all the associated requirements for tool steel, precision inserts and cavities, ejector pins, correctly sized gating and the like, necessary for the process to work satisfactorily. A major factor in injection moulding is that the tooling cost should be properly written off over the required length of production run, bearing in mind the overriding requirement for a good product. Recently, Walshe & Lowe (1984) have provided information on this point.

6.9   Some aspects of plastics selection and design

In an ideal world, a person designing objects and selecting plastic materials from which to make them would have had an academic and parallel practical education involving all the topics listed in Table 6.11. In reality he is

Table 6.10. *Assembly methods for thermoset plastics*

Assembly methods	Adhesive bonding	Induction bonding	Mechanical fastening
Thermosetting resins			
Alkyds	X		
Allyl diglycol carbonate	X		
Diallyl phthalate	X	X	
Epoxy resins	X		X
Melamines	X		
Phenolics	X		X
Polybutadienes	X		
Unsaturated polyesters	X		X
Silicones	X	X	
Ureas	X		
Polyurethanes	X		X

unlikely to be familiar with more than a few of these concepts, and would hold some preconceived ideas through being a practical user of the materials in private life. Beck (1980) provides, very comprehensively, a checklist of all the design rules which should be followed to allow for the peculiarities of plastics as a class. For example, one must take cognisance of the large thermal contraction occurring on solidification at the end of a thermoplastic moulding process; this imposes a requirement for adequate tool tapers, and elimination of ultra-thick sections and very large section changes in the moulding itself. It is when the actual material class itself must be selected that the designer tends to stand alone, and hence to need the most help. Such help can range from lists of available materials showing some of the end uses, Knight (1983*b*), to tabular checklists reminding the designer of the requirements to be met, followed by bar-chart presentation of the main property parameters, BNCM (1984) to card indexes. Of these a recent offering is the *Engineering Thermoplastics List*, which comprises 36 pages of detailed data sheets on nearly all the current plastics and their variants, with main areas of application, chemical name, manufacturer's trade name, physical properties and price guides being provided for each, Anon. (1984*a*).

Perhaps the most sophisticated route in future will be a selection system based on computer technology. For example, the system PLASCAMS 220 is

Table 6.11. *An idealised plastics education syllabus for a designer*

Simple organic chemistry
Molecules and the nature of plastics
The main families of plastics, including the concepts of thermoplasts, thermosets and elastomers
Composite plastics, fillers and foams
Trade names and major suppliers
Identification of plastics, with a reference test kit to assess the effects of fire and solvents
Specifics of the plastics families
Mechanical properties as a function of temperature, time and strain rate
Environmental effects of light, heat, chemicals, environmental stress cracking
Flammability and the hazards of toxic smoke
Methods of joining, fastening and fabrication into products
Available surface finishes
Design pointers for plastics
Cost comparisons with other materials per unit of volume, height and strength
Maintenance and repair. Service life expectancy
Reclamation or disposal
Test methods, standards and codes of practice
Some important organisations in the plastics field
Bibliography of useful references

Table 6.12. Properties of typical foamed polymers

Material	Specific gravity	Tensile strength (N/mm²)	Izod impact (J/cm²)	Elongation at break (%)	Compressive strength (N/mm²)	Heat deflection temperature (°C)	Thermal conductivity (W/m K)
Rigid Foams							
Polyphenylene oxide	0.8	23.0	0.55	15.0	35.0	96	0.124
Polycarbonate	0.8	37.7	0.74	4.0	35.0	128	0.151
Epoxy resin	0.78				1.0		0.7
Isocyanurate	0.032	0.3			0.2		0.1
Polyether	0.08						
Polystyrene	0.17				1.5		0.65
Polystyrene	1.04	52.7	0.26	3.5		101	
Polyurethane	0.11	0.4			0.6		0.3
ABS	0.86–1.1	13.0–27.0	0.4–0.95		31.0	72	
Acetal	1.130	74.0				153	
Nylon 6/6	0.97	94.0		4.1		255	
Polybutylene terephthalate	1.1	70.0	0.38	1.3	79.0	207	
Polyimide		120.0				277	
Polysulphone	0.87	35.0		3.5	35.0	177	
Polyvinylchloride	0.6	10.0		370.0			

available in the form of floppy discs which can be run on a variety of micro-computers and was jointly developed by the RAPRA Technology Ltd and the Lucas Company in the UK. The selection system contains data on over 220 generic classes of plastic, covering the full range of thermoplastics, thermosets, composites and thermoplastic elastomers. The computer can offer a choice of materials to meet the product specification parameters fed into it. A major problem with this type of system lies not in the data held in the computer but with the designer, who will need to learn afresh how properly to write specifica-tions. These should be prepared so that only a reasonably small number of can-didate materials are chosen for each operation; production of these finely wrought specifications demands practice.

A somewhat esoteric subject is the question of design in structural foam, which is becoming widely used for making housings for business machines and doors for buildings. Some typical properties are offered in Table 6.12. Foamed articles can have many built-in mechanical features to simplify internal assembly arrangements and at the same time avoid some of the problems inherent in injection moulding very large parts in solid plastic. A useful text from Colbert (1980) discusses the beneficial absence of sink marks and moulded-in stresses in foams, with the ability to mould zero degree draught in limited areas. Colbert stresses the theme of good design affecting the economics of moulding in relation to the size of the projected production run. This is equally important whether thermoplastic or thermoset structural foams have been chosen.

6.10   Some recent developments in polymers

The designer must always be receptive to new ideas for exploitation in new products. At the time of writing, one or two threads of recent development have surfaced which should have enjoyed considerable exploitation by the end of this century. For example, electrically conducting polymers are now the focus of intense interest. Although in the past it has, of course, always been possible to incorporate conductive fillers into otherwise insulating polymers for static charge dissipation and other purposes, emphasis in the 1980s is on building intrinsic electronic conduction into a polymer molecule so that the material is suitable for use in batteries and electrical contacts. A promising con-ductive plastic is polyacetylene, made conductive by controlled doping. Clarke (1980) shows how films grown directly on an electrode material by electro-chemical oxidation can have a doping gradient built up across them. Unfor-tunately, polyacetylene dopants tend to be rather reactive towards moisture and air, so their applications are likely to be limited until displaced by more stable materials. More recently Pye (1983a) offers a survey of work on highly con-ducting polymers since 1971, discusses conduction processes in polyacetylene and focuses on the use of this and other materials in batteries with conductive plastic electrodes. Even more recently, Young (1984) reviews developments indicating the commercial possibility of a plastic battery.

Polymers other than polyacetylene exhibit conductive and, indeed, other spectacular effects. Conductive polymers based on the reaction of various bipyridenes with ruthenium exist in at least eight discrete oxidation states which are stable for long periods in the absence of oxygen and water. Their colours include orange, purple, blue, green – blue, brown, rust and cherry red, and, if incorporated into polymers forming the surfaces of electrodes, may offer the basis for new types of display devices, Worthy (1983).

Unlike polyacetylene, polypyrrol shows distinct promise for applications where exposure to air is inevitable. Although the material has only about 20 % of the maximum conductivity achievable with polyacetylene, it has much better resistance to oxidation and is usefully available in flexible film of several thicknesses, British Plastics and Rubber (1983).

Fabrics made from plastics are, of course, as old as the first synthetic rayon. Even the melded (non-woven) fabrics, which came to the market place some years ago, have now achieved respectability and, among other uses, are employed for stabilising the foundations of roads or buildings built upon sandy or marshy ground. No less successful, but not quite so well known, is the range of fabrics and fibres made by the expansion of polytetrafluoroethylene (PTFE) without the use of a blowing agent. PTFE is first formed into a rod by extrusion, being mixed with the lubricants normally used for this purpose, but left unsintered. While being heated to less than the transition temperature of 327 °C, it is stretched at a high rate of strain and then forms a porous fibre which, because of molecular orientation, is considerably stronger than when the material has been sintered without previous stretching. The product can have a porosity as great as 96 %, but more interestingly has a tensile strength more than thrice that of the solid polymer, even when the porosity is 90 %. This means that the tensile strength of the non-porous matrix fibre must be approximately 700 N/mm^2 — a formidable achievement. Hinds (1983 – 84) provides further details. This development brings nearer the prospect that a similar process applied to other materials may produce other 'super-fibres'. Such fibres were presaged by Coates & Ward (1981), who show how the drawing of polypropylene at a temperature below the melting point yields a product considerably in advance of those from conventional drawing, hydrostatic or ram extrusion processes. The polymer forms rods with very high stiffness; the process has the useful characteristic that an increasing speed of drawing not only reduces the unit cost but produces a material of even higher stiffness.

7

Properties of composites

As with ceramics, the designer will often be forced to work with composites by designing an object which has to be made by an outside specialist. He will seldom be in a position to purchase ready-made stock shapes from which his brainchild can be fabricated. Accordingly, much of this chapter will be devoted to showing the designer a general view of composites as a class of materials so that he will be able to pose the correct questions to his advisers and suppliers, and perhaps even to understand their answers.

7.1 Definition of composite types

The term 'composite' covers a very wide variety of forms. Thus a concrete path can be considered a composite, as may a printed circuit board used in a television set. However, for the purposes of the present discussion, emphasis will be placed on composite materials to be used in engineering but not in building or electronics. Composite constructions can also readily be dismissed. Thus, for example, a sheet of plastic covered with a metal film for decorative or strength purposes, or a concrete – steel composite floor used in a building, should really be thought of as composite structures and not composite materials. By the latter we mean essentially a material in which there is a continuous phase, usually representing the greater proportion of the whole by weight or volume, containing dispersed discrete islands of a second phase. Thus we are considering a plastic, metal, ceramic or glass mass which could be free standing, but which has properties modified by incorporation of particles, fibres, or woven fabrics. The wide range of disperse phase materials that can be incorporated, and the ways in which they can be distributed randomly or with a preferred orientation, offers enormous choice in optimising any given composite and in achieving any combination of performance parameters required.

Some of the possibilities are illustrated in Figure 7.1. Particulate matter dispersed in the matrix could range in size from the very small dimensions of a minor proportion of metal phase precipitated from a heat-treatable alloy, up to quite large metal or glass flakes or microspheres. Similarly, glass fibres incorporated may range nominally from two millimetres long up to ten millimetres, and this dimension can be extended considerably if chopped fibre strand mat or whole pieces of cloth are included. These one- and two-dimensional reinforcements are converted into, effectively, three-dimensional forms by filament winding of a reinforcing fibre on a mandrel to manufacture a tube, or by the inter-reaction of organic materials to produce interpenetrating polymer networks.

As a general rule, materials containing very small proportions of an added disperse phase as, for example, the incorporation of flame retardants or smoke suppressants into a plastic, or the adventitious content of some impurity in a metal, are not considered as composite materials. The term 'composite' in this chapter implies the deliberate addition of from a few percent to perhaps 60 % of a disperse phase, which is usually harder than the matrix and which is added to achieve some net property improvement. In any event, the material has to be so formulated that good bonding is achieved between the reinforcing material and the matrix; this may be achieved by natural serendipity or, more reliably, by the use of coupling agents which act as a primer on the disperse phase surface.

Table 7.1 summarises materials commonly used as reinforcements and fillers, provides some idea of the types of matrix with which they are usually associated, and indicates some general applications for them.

Fig. 7.1. Structure of composites. (Copyright: 1985 Standard Telephones and Cables plc.)

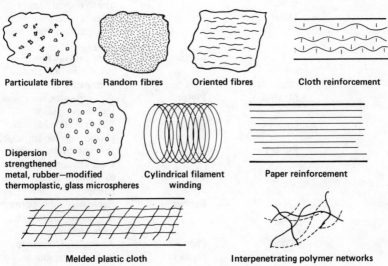

Particulate fibres   Random fibres   Oriented fibres   Cloth reinforcement

Dispersion
strengthened
metal, rubber—modified   Cylindrical filament   Paper reinforcement
thermoplastic, glass microspheres   winding

Melded plastic cloth   Interpenetrating polymer networks

Table 7.1. *Materials in composites*

Reinforcement or filler		Matrix	Some Uses
Fibres	Carbon	High-strength plastics	Sports goods, aerospace
	Ceramic	Metal	High strength electrical conductors
	Glass	Concrete Plaster All forms of plastics	Walls May be moulded or filament wound
	Metal	Plastic	To be made conductive for electrical shielding
	Plastic	Plastic	Mouldings, cable strength members
Molecules	Polymers	Plastic	These 'molecular fibres' appear in oriented thermo-plastics, block and graft copolymers, especially for thermoplastic elastomers, and in interpenetrating polymer networks
Particles	Carbon black	Rubber	Filler and UV light protection
	Ceramic	Plastic	To modify dielectric constant
	Glass	Plastic	Glass microspheres are incorporated to fill space and to reduce density
	Metal	Metals	Deposited as fine grains in precipitation- hardened alloys
		Plastic	Used in flake form for decorative or electrical purposes
	Mineral	Plastic	A cheap space filler which increases rigidity
	Organic	Plastic	Starch is added to make plastic bags biodegradable
	Rubber	Plastic	Precipitated by chemical means in complex plastics to improve impact strength
		Rubber	Reclaimed rubber crumb is used in vehicle tyre manufacture
	Wood	Phenolic plastics	As a reinforcement to increase strength

Table 7.1. (*cont.*)

Reinforcement or filler		Matrix	Some Uses
Sheet	Glass cloth	Polyesters	Used to make reinforced panels for buildings and vehicles
	Melded plastic mesh	Plastic	A toughening interlayer for horticultural and roofing light weight sheets
	Natural fibre cloth	Thermosetting resin	To make strong laminates for use as bearings and in electrical applications as an insulator

7.2 Reinforcements in common use

If the purpose of making a composite is to obtain a material stronger or better in some ways than the original virgin matrix material, it follows that any added reinforcement should be stronger and stiffer than the matrix, and should be able to modify the most likely failure mechanism in a favourable way. A reinforcement with high strength and high stiffness probably has little or no ductility and thus will behave in a brittle manner, the effect of which is minimised if the material is used as filaments or fibres. All filaments or fibres used must have an adequate length to width ratio, so that they can effectively overlap each other in at least one axis of a properly constituted composite, so enabling their strength to be available to all parts of the product. Early fibres used were naturally occurring, ones such as silk, cotton, horsehair, jute and asbestos. However, in relatively recent years, the joint development of high-performance plastics with the ready availability of high-strength glass fibre, led to a considerable period of development during the Second World War, the momentum of which has increased to the present day. So high-performance composites always tend to depend upon fibre reinforcement, although of course particulate materials have their place.

Glass is the most common fibre-reinforcing material, available in a variety of types able to withstand those different environments encountered within the matrix material during service. Glass is unique among other fibres insofar as it is essentially available in unlimited lengths, enabling cloth to be manufactured from it economically. Many other fibres, owing to their manufacturing process, have a length limitation, although they can be stranded together to form a tow from which cloth can be manufactured.

Carbon fibres have received considerable publicity in the past few years because of their potential for making composites of unusually high stiffness.

These fibres were first produced by pyrolysis of a rayon precursor, but latterly are made by heat treatment of polyacrylonitrile fibres or from pitch. Polymer fibres and inorganic fibres are also made and used widely, a range of those in general use being shown in Table 7.2.

Although glass fibres can be obtained in long lengths, in reality they are frequently used for reinforcing polymeric materials as short chopped lengths of ten millimetres or less. Unfortunately, during conventional processing of these materials through the extrusion or injection moulding routes, these fibres can be badly broken up and may have a final length of only two millimetres in the finished article. Recently, some new technologies have been developed which enable ten-millimetre lengths of fibre to be preserved in moulding granules which, with care, can then be fabricated into objects significantly stronger than the current norm. Even more strength is achieved if very tough thermoplastic material, like polyether etherketone, is used as the matrix. Norman (1981) reports a significant study of aromatic polyamide fibres as a possible replacement for asbestos and suggests advantages and limitations for the material compared with other fibrous reinforcements in a resin binder.

Table 7.2. *Reinforcing fibres in common use*

Alumina	Mineral wool
Asbestos and its replacements	Natural fibres
Beryllium	Polyaramide
Boron	Polyester
Boron nitride	Rayon
Brass	Silicon
Carbon	Silicon carbide
Glass	Steel
Tungsten	

Table 7.3. *Particulate fillers commonly used in composites*

Alumina	Metal powder and flake
Antimony oxide	Molybdenum disulphide
Barium ferrite	Polymers
Barium sulphate	Potassium titanate
Calcium carbonate	Silica
Carbon	Silicates, e.g. talc, mica
Glass	Starch
Hydrated alumina	Titanium dioxide
Magnesium carbonate	Woodflour
	Zinc oxide

The number of types of particulate filler used in composites is legion, but often their exact nature is blurred by commercial modification and conditions of secrecy. However, Table 7.3 lists some of those most commonly used in polymer composites. Other particulate fillers might be formed in situ, for example in precipitation hardened metal alloys and in those which are dispersion hardened. These fillers will often offer specific technical functions to a polymer. For example, barium ferrite imparts magnetic properties, barium sulphate increases bulk density, hydrated alumina provides some resistance to spread of fire, titanium dioxide confers opacity, and so on.

7.3   Advantages and limitations of composites

It can generally be claimed that fibre-based composite materials offer good potential for achieving high structural efficiency, coupled with weight-saving in products, fuel efficiency in manufacture, and cost effectiveness during service life. Conversely, it can be pointed out that special problems can arise from the use of composites due to the extreme anisotropy of some of them, from the fact that the strength of constituent fibres is intrinsically variable, and because test methods for measuring composite performance need special consideration if they are to provide meaningful values.

Some of the advantages, in terms of high strength to weight ratio and high stiffness-to-weight ratio, can be seen in Figure 7.2, which shows that some composites can outperform steel and aluminium in their ordinary forms. However, it should be remembered that these values do not include acceptable

Fig. 7.2. Tensile properties of some composites and unreinforced materials. (Copyright: 1985 Standard Telephones and Cables plc.)

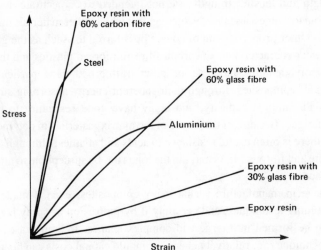

safety factors, which may require a value of three. If bonding to the matrix is good, then fibres augment mechanical strength by accepting strain transferred from the matrix, which otherwise would break. This occurs until catastrophic debonding occurs. Particularly effective here are combinations of fibres with polymer matrices; they often complement one another's properties, yielding products with acceptable toughness, reduced thermal expansion, low ductility and high modulus. Apart from these combinations, most composites are expected to combine the good properties of two or more materials more cheaply than could be achieved with any other single material. Composites offer the option of achieving maximum strength in pre-determined axes by aligning fibres, or by the use of two-dimensional woven cloth reinforcement. On the other hand, particulate solid fillers in polymeric matrices tend to yield isotropic properties in composites, which is useful because filler particles act as crack propagation stoppers, thus increasing strength under adverse environments. However, at the high rates of strain occurring upon impact, solid fillers in general do not increase the strength, although rubbery particles incorporated on the molecular scale in acrylonitrile – butadiene – styrene polymers do have this useful property.

As a further advantage, composites make effective use of some materials otherwise unable to stand alone, such as mineral fibres or wood flour. When incorporated into polymers, in particular those such as unsaturated polyesters or phenolics, particles can reduce manufacturing shrinkage and yield a more usable product. In service, zero thermal expansion coefficients can be achieved by suitable choice of starting materials.

Some selected properties of a limited range of polymer composites are provided in Tables 7.4, 7.5, and 7.6. The values offered for general properties, impact strength and tensile strength, are not intended to be accurate design parameters, but to suggest a scale of values against which other materials might be compared. Other properties can of course be referred to, such as the good friction and wear characteristics of carbon-fibre-reinforced plastics and those filled with molybdenum disulphide or polytetrafluoroethylene particulate fillers, while all composites, through their inherent energy-absorbing struc-ture, have high damping capacity, and many have good resistance to heat transfer and fatigue. Because of the nature of the matrix materials in polymeric composites, there is often useful resistance to acids and alkalies, although cer-tain particulate and fibrous fillers may, in the long run, be susceptible to attack near the surface.

An advantage in manufacture is that many composites, even if made by injection moulding techniques, and certainly if by pultrusion or spray lay-up methods, can be formed into large and complex shapes. In some cases the fabrication techniques are relatively cheap, as mould capital costs can be low.

Table 7.4. *Properties of some reinforced plastics and of their parent matrices*

Material	Specific gravity	Tensile strength (MN/m²)	Flexural strength (MN/m²)	Flexural modulus (MN/m²)	Impact strength (notched) (J/cm)	Coefficient of linear expansion (×10⁻⁶/°C)	Heat distortion temperature (°C)
Polypropylene (glass reinforced)		54–83	68	202	1.86	40	140
Polypropylene (base polymer)	1.070	30	47	101	0.80	100	60
Polystyrene (glass reinforced)		74	95	67	1.60	35	110
Polystyrene (base polymer)	1.060	61	74	34	0.16	80	95
Polyamide (glass reinforced)		135	189	67	1.86	25	220
Polyamide (base polymer)	1.240	74		17	1.06	80	80
Polyester laminate		95	175	7	5.3	25	
Polyester (54 % w/w glass rovings)	1.9	80			12	18–35	175
Polyester (45 % w/w woven glass rovings)	1.6	25			12.5	15	175
DMC Polyester (10–40 % w/w glass fibre)	1.8–2.0		39–137			24.34	230
SMC Polyester (20–35 % w/w glass fibre)	1.8–1.85	75	137–206			18–33	175
Epoxy resin (bi-directional glass fibre)	1.5–2.1	240	343–617				

However, the fabrication processes are usually more labour-intensive than most in the plastics industry. Glass-reinforced plastics incur costs of finished mouldings which are comparable with those of similar shaped articles made of metals, a fact which has not escaped the manufacturers of small numbers of specialist motor cars. Lastly, it should be pointed out that the properties of com-

Table 7.5. *Impact strength of some polymer matrices and their composites*

Material	Impact strength (J/cm)
Epoxy resin	0.1 − 0.6
Epoxy resin + 20 % glass fibre	5 − 15
Nylon 6	0.5
Nylon 6 + 20 % glass fibre	0.7
Nylon 6 + 40 % glass fibre	1.6
Phenolic resin	0.15
Phenolic resin with cotton cloth	0.4
Polyester/styrene, cross-linked	0.1
Polyester with glass cloth	2.3
Polyphenylene sulphide	0.15
Polyphenylene sulphide + 40 % glass fibre	0.4
Polystyrene	0.14
Polystyrene + 30 % glass fibre	0.58

Note that some materials, including polyacetal and polyethylene, show a reduced impact strength when incorporating glass fibres.

Table 7.6. *Tensile strength of some composites*

Material	Tensile strength (MN/m^2)
Aluminium	70
Epoxy resin + 60 % boron fibre	900
Epoxy resin + 50 % carbon fibre	580
Epoxy resin + 60 % glass fibre	2000
Nylon + 35 % glass fibre	185
Polyester + 30 % chopped glass mat	103
Polyester + 40 % glass fabric	320
Polyester + 80 % filament wound glass fibre	1000
Polyimide + 30 % glass fibre	90
Polyphenylene sulphide + 40 % glass fibre	160
Polysulphone + 40 % glass fibre	120
Polytetrafluoroethylene + 20 % glass fibre	78
Steel	200

posites can often be varied and tailored according to the design requirements to be met.

On the other side of the coin, some limitations in the use of composites are readily identified. For example, the stiffness of glass-reinforced plastic composites is low compared with metals, although it can be improved by altering the geometry, by producing sandwich structures or by using special fibre reinforcement. Many composites involving the use of cloth or paper reinforcement are susceptible to splitting and delamination; cracking can occur if the primer bonding on the surface of fibres is not adequate for adhesion to the matrix. Stress then causes formation of cracks into which moisture can permeate. Breakdown of reinforcing fibre length can often occur during processing, reducing strength below what was calculated as theoretically possible.

Although a composite material may have considerable strength in one axis, this may be at the expense of reasonable strength in another. For example, parallel lay-up of glass fibres in an epoxy resin matrix will produce very powerful leaf springs in the axial direction. However, in a direction normal to the axis of the glass fibres, the material is very weak and certainly has no engineering functions unless additional support is used.

Composites may be degraded by water and solvents beyond the action on the polymer matrix alone. This is particularly the case when liquids can permeate through tracks and paths caused by the phase discontinuity of the reinforcing material. Carbon-fibre-based materials are electrolytically active, which means that when in contact with adjacent metals they may accelerate corrosion. Some metal fillers, on the other hand, can accelerate thermal degradation of organic polymers through surface oxidation and subsequent catalytic action.

Although filled composites and, to some extent, reinforced composites can often command prices little different from those of the matrix material, there are instances of some very high materials costs for advanced-type fibres such as carbon and boron, limiting present use to a narrow market.

The nature of the occasionally complex fabrication techniques can lead to a variation between components which might be unacceptable for their performance in critical applications. Another problem with fabrication lies in joining techniques; they may be compatible with the manufacturing process in the case of polyester – glass cloth, and lay-up of glass-reinforced plastics, but may be very difficult for other materials which cannot easily be adhesively bonded. Mechanical joining generally depends on boring holes which produce stress-raising areas on their periphery, with the potential for later catastrophic failure.

A general problem, obvious from a scrutiny of Tables 7.4, 7.5 and 7.6 is that of providing the designer with specific information relevant to identifiable composite materials on which he can base his calculations. Although it is possible to say that the impact strength of a nylon 6 containing 20 % glass fibre by weight or volume is 0.7 J/cm, this begs the questions of which of the many available grades of nylon 6 is to be used, the way in which the glass fibres should be dispersed isotropically or anisotropically, the optimum nominal glass-fibre length and the effectiveness of coupling of the surface to nylon. In this context, performance values have been deliberately set out in the tables to set the scene for those improvements which might be met by incorporation of fillers or reinforcing materials into what, in other circumstances, might be a self-supporting polymer matrix.

In some cases, commercial suppliers are able to provide very relevant and specific information. For example, E.I. DuPont de Nemours Inc. provides a data book dealing with its Kevlar 49 (registered trade-mark) polyaramid fibre used for filament winding of composites. The book deals with the general properties of the fibres themselves, then discusses the specific types needed for filament winding, and goes on through environmental performance to the actual process of filament winding and the materials used in it. Considerable information is then offered on the properties of laminates regarding their creep resistance to high loads and exposure to temperature. Such information should be sought by the designer from his materials suppliers, and carefully cherished when he finds it.

7.4   Design considerations

In engineering applications, the designer is generally more eager to achieve a greater strength for his product through use of a composite material than through selecting a still single, economic, other material. In a consumer-oriented world this may be compounded by a need to load the base material with some benign, cheap filler to reduce the materials cost part of his product equation. Particulate fillers are useful for that purpose, but most of the present discussion will centre on the positive benefits of using fibres and sheets as components of a composite. Usually the final strength of a composite component depends very much upon that conferred by the presence of the fibres which are usually of high stiffness and high tensile strength. However, these fibres must be bonded together in some useable, rigid form to perform their function adequately. This bonding is achieved by using a matrix which may be a polymer, ceramic, glass or a metal, which itself exhibits properties chosen to be complementary to those inherent in the fibres themselves, or perhaps to provide some property which the fibres do not have. A typical example is of a matrix material chosen to provide toughness, i.e. to resist crack propagation between fibres and matrix, leaving the fibres themselves to provide high strength and stiffness. In other cases, the matrix might be used to hold together small pieces of particulate material which alone perform the final function. One example is cermet cutting tools for metals, in which small grains of cutting ceramic are held together by a metal which wets and binds them but which does not directly contribute to the technical performance of the tool, Anon. (1983*b*). Again, in automotive tyres, steel or other fibres are used to provide long life and geometrical stability, while the rubber content binds the whole together and provides the necessary degree of flexibility, resilience and grip.

To achieve maximum strength in a given direction in a fibrous composite, the reinforcing fibres should be aligned parallel to that direction. However, this arrangement leads to strength anisotropy and thus probably inadequate strength in other directions. More usually fibres are laid in alternate and controlled directions, as happens with woodgrain in plywood sheets, or perhaps are used in shorter lengths and aligned randomly in a less maximised design option. However, when a large volume fraction of fibre is to be incorporated, the random approach is not so useful because of the lower packing efficiency that can be achieved. One way of achieving high packing density is to use fibres made into cloth (as in the case of glass-cloth reinforcement for large panels and structures) or by the use of a filament winding technique which, however, is mostly suitable only for tubular or hollow bodies such as pressure vessels and pipes.

High strengths in a composite are only achieved if the fibres are able to accept large elastic strains, which themselves might generate unacceptably high deflection values. Because of this, the properties of composites should be well-known before there is any thought of substituting them for an established material in an existing product, as the ultimate performance may be unacceptably modified by the tendency of high deflection in some composites.

Something has already been said about the problem of toughness in composites; the designer has to achieve a compromise between strength and toughness in his construction. High fracture toughness depends upon having a high aspect ratio for the fibre reinforcement, coupled with weak interfaces along the length of fibrous or sheet reinforcement with the matrix material. On the other hand high strength in tension requires a low aspect ratio and a strong interfacial bond between the fibres and the matrix. These are subjects where, although the designer can obtain some information from his own theoretical sources and those of suppliers, he is very dependent in the long run on the degree of workmanship put into products made in areas outside his direct control. For example, the number of air bubbles or inclusions in a wet lay-up process could very quickly become unacceptable without good control by the process operator.

In a composite in which the particular fibre or other individual reinforcing material unit is randomly oriented throughout the bulk, we have a material which is essentially isotropic, that is, it has the same properties in any direction. However, in all other composites, for example use of oriented fibres or impregnated cloth or paper, more or less anisotropic properties follow because the reinforcing geometrically alters the strength in a preferred direction. The elastic response of isotropic materials is specified by two independent constants, the familiar Young's modulus, and the, perhaps less familiar, Poisson's ratio, which is the ratio of the strain in the direction of the principal applied stress to that produced in the transverse direction. However, anisotropic composites require knowledge of at least two Young's moduli and two Poisson's ratios' elastic responses to be estimated and specified. Such properties can be measured with difficulty, bearing in mind the complexity caused by building up laminated layers of composite sheets from successively deposited layers of cloth oriented in different directions to each other.

Failure to take proper account of variations in elastic response can lead to twisting of composites when a simple tensile load is applied. Although, in some instances proper analyses can be made, the normal commercial designer may at best receive only upper and lower property value information. Some properties, such as thermal and electrical conductivities, and the elastic properties, can be calculated from models based upon an idealised product such as a filament-wound tube, and this works well enough provided that the model closely approximates the real product geometry. However, other properties such as failure strength and fracture toughness are not easy to calculate and can only be approximated at best.

When the reinforcement of a composite consists of parallel fibres in one direction only, the rule of mixtures allows calculation of some material parameters such as heat capacity, density and Young's modulus. Some approximations can be made for the various Poisson's ratios.

The foregoing comments will probably make the reader feel that he is entering a minefield of complexity, difficulty and black art. This is probably quite true, so the average practitioner is well advised to rely upon the experts who have gone before and who have set out techniques for designing piece parts in more understandable form. One such, dealing with discontinuous glass-fibre-reinforced thermoplastics, is offered by the US Department of the Army (1981), and provides guidance on the shrinkage, dimensional tolerances and warpage of parts, and the proper use of drafts, undercuts, ribs, bosses, holes, fillets, radii, wall thickness, and weld and parting lines. This useful text deals with joint design, the various methods of producing adhesive or thermal bonds, and discusses threads, surface finishes and safety factors.

Perhaps easier to understand, and nearer to home, is the occasional publication on design data for glass-fibre composites offered by Quinn (1977). This provides mathematical design rules covering glass fibre materials, and product information such as prediction of thickness for laying up polyester and glass composites, calculation of density and many general examples of calculation of composite strength from a knowledge of the construction. A useful section suggests property prediction, for example from graphs of tensile modulus and tensile strength, as a function of glass content.

Despite the apparent difficulties in understanding the theory of composites, successful designs using them are quite common, as some of the examples of Table 7.7 will show.

7.5   Production techniques

Despite the wide range of composites available, in this section we will consider only the general cases of composites with plastic or metal matrices.

Methods chosen to fabricate composite materials and structures depend upon the type of reinforcement and matrix chosen, the performance level required, the size of the production run and the shape of the article to be made. Whereas short fibres cannot normally be positioned exactly within the body of a material, except by the orientation achievable through extrusion or injection moulding, continuous fibres and filaments are controllable and tend to be used for the manufacture of products at the very highest integrity, for example in the aerospace industry. By laying parallel arrays of fibres with a machine, tapes can be formed which are overlapped in a criss – cross manner to form quite complex shapes, either on flat surfaces or by winding on a mandrel. This technique works for relatively large diameter single filament materials such as reinforcing wires. When, however, the filaments are very fine, as in glass or carbon fibres, bundles of many thousands of the fibres are wrapped together to form a tow or roving, and these are then treated as single strands.

For manufacturing polymer matrix composites, fibre bundles or single fibres can be incorporated into a semi-processed polymer to form a dough-like ribbon or sheet, known as a 'pre-preg', in which liquid resin is incorporated into the fibre rovings. The impregnated material is processed and dusted so that it can be handled as a soft, dry entity, which is then laid in a mould or onto a mandrel

Table 7.7. *Some successful applications of composites*

Cemented carbides for cutting tools
Consumer goods
Dental cements
Electrical parts
Moulded plastic products
Panels for buildings and vehicles
Plain journal bearings
Plastic pipes
Sports goods
Vehicle tyres

surface, sometimes being partially cured while progressive build-up occurs. Subsequent oven treatment then fully cures the resin matrix and provides what can be a very complex shape ready for use.

The process of pultrusion occurs when continuous fibres, in the form of rovings impregnated with resin, are formed to the desired overall profile by a die and then pulled through it and simultaneously cross-linked to form a composite bar, rod or channel having the required strength in the linear direction.

When products of a lower order of strength are required, as for cladding panels for vehicles and buildings, then pre-preg sheets can be made using chopped glass or other fibres, which have the advantage of being mouldable into doubly curved shapes without serious buckling. However, the lack of really long fibres in it means that the ultimate strength is reduced although probably still acceptable for most applications. Where very large shapes such as boat hulls, or, alternatively, a small production volume, are needed, then the spray-up technique is used. Special spray guns accept continuous filament roving which is chopped in the gun to produce short fibres, and mixed and dispensed simultaneously with a liquid resin to build up a composite material laid down onto a mould surface. After curing with chemical catalysts or ultraviolet light, the material is sufficiently form-stable for final oven curing.

Thermoplastic-reinforced materials, including those containing particulate fillers or short fibre lengths, can be processed by the normal methods used for pure polymers, that is, by injection or transfer moulding, compression or extrusion, bearing in mind that these materials need a limited reinforcement content to permit acceptable flow conditions. In some of these processes reinforcing

Table 7.8. *Production routes for polymer composites*

For solid polymers with incorporated particles or fibres	For liquid or pasty cross-linkable resins mixed with fibres or cloth
Calendering	Casting
Centrifugal moulding	Compression moulding
Extrusion	Dough moulding
Injection moulding	Filament winding
Transfer moulding	Hand lay-up
	Lamination
	Prepreg pressing
	Pultrusion
	Reinforced reaction injection moulding
	Sheet moulding
	Spray lay-up
	Vacuum forming of sheet
	Vacuum impregnation

materials, at first added randomly, become partly oriented by the flow distribution of the material during moulding and this can sometimes be exploited to confer increased strength in preferred directions. Increased wear of machines, tools and moulds often occurs when reinforced materials are processed.

The general processes for making polymer composites are listed in Table 7.8, but a better exposition, particularly of the different methods' advantages and limitations, is provided by Pye (1979).

Fibrous composites with a metal matrix can be made by a number of routes. For example, bundles of parallel fibres or rovings, whether of alumina, glass, or perhaps metallic fibres like tungsten, can be infiltrated with molten matrix material to yield a mass which can then be rolled, extruded or machined.

A simpler method is the directional solidification of alloys, in which the reinforcing phase might be fibrous or laminar. Some combinations use tantalum with tantalum carbide or niobium with niobium carbide, but nickel and cobalt-based eutectic alloys are more widely used as matrix materials. Directional solidification involves a casting process in which the required shape is directly produced and needs only finishing machining of significant surfaces.

Powder metallurgy with its mixing, pressing and sintering, or direct hot pressing stages, is widely used to provide a required shape direct, or to generate a semi-finished composite for subsequent hot working. Some combinations include nickel with silicon nitride whiskers, cobalt with tungsten fibres and aluminium with steel reinforcement.

Sometimes the matrix can be deposited from the liquid phase; thus fibres of alumina or tungsten may be electroplated with nickel, or nickel can be deposited by a chemical vapour deposition process onto the fibres, followed by hot pressing to compact the mass which is then machined to final dimensions.

Lastly, mention is made of a lamination process in which aluminium and steel, for example, can be hot pressed together in the form of alternate layers of matrix sheet and dispersed fibre. These hot-pressed sheets are subsequently repressed into the required shapes.

7.6   Reinforced plastics

Reinforced plastics probably comprise the greatest bulk of composites used in product engineering applications. The two main types of reinforced plastic are those in which a normally thermoplastic polymer or elastomer is filled or reinforced with particles or fibres, and those in which a liquid precursor resin is used to impregnate a reinforcing mat or roving of fibres, being subsequently cured into a solid for final use. The total use of reinforced plastics in Europe has risen from 300 000 tonnes in 1970 to 600 000 tonnes in 1980, these figures being exceeded by the American market. Their effect has been to offer considerable freedom to the designer. Structures can be built up from fibres, tapes and cloths with plastic matrices to any desired thickness. By varying the type and dimensions of the fibre, the orientation and volume fraction in the composite, as well as by varying the matrix material, multi-directional laminates with a wide range of stiffness, strength and resistance to environments can be produced. Such materials can also be built into cellular or honeycomb structures to yield the additional advantage of weight saving.

In general, fibre-reinforced thermoplastics or thermoplastic elastomers show some degree of orientation of short-length fibres during production of the finished part, yielding both advantages and limitations to the strength in different orientations. A disadvantage inherent in the commonly applied injection moulding process, or in extrusion, is that fibres initially chosen for dimensional reasons are broken during processing, and their shorter progeny in the final product cannot achieve such high final-strength values as were expected. New techniques are, however, being announced to maintain reasonable fibre lengths; 'reasonable' in this context being ten millimetres or more.

The strength and stiffness of the fibres and the retention of these properties during the composite's service life, are the reason for adopting composites in the first place. The resin matrix performs by spacing the fibres, preventing their mutual abrasion, and acts as a barrier to attack by the outside environment. The matrix also holds the fibres geometrically within the required shape.

Resins such as phenolics, alkyds, unsaturated polyesters, epoxides and polyimide thermosetting resins have all been used as matrices, generally mixed with glass fibre or glass cloth, and often partly cured to form pre-pregs which are subsequently shaped and heat-cured into the final composite form.

Although considerable reference has been made in earlier paragraphs to the use of glass fibres as a reinforcing material, largely because they are effective and inexpensive when used in almost all common thermoplastics and thermo-sets, other fibres commercially used include carbon, cellulose and inorganic materials such as ceramics, silicon carbides, silicon nitride, and boron. Brass-plated steel wire and potassium titanate fibres are used; the latter act as a pigment and also confer electroplateability on the matrix material. Synthetic organic fibres are used, including polyamides, polyaramids, polyesters and polyvinyl alcohol.

Finally, mention must be made of the ubiquitous GRP (glass-reinforced polyester) which is a high-strength material made from about three parts of polyester resin and one of glass fibre in the form of rovings or woven fabric. Pigments and inert fillers are incorporated, so that after curing a high-strength material is obtained, having easily been formed into complex shapes. Because of the high bulk cost, this material is mostly used in thin sections for decorative cladding on buildings and vehicles, of which a recent example is the front of the

British Rail APV train locomotive. The material is generally impervious to moisture and has a virtually permanent colour and surface finish. Like other plastics, there is likely to be long-term deformation under continuous stress, but provided a properly designed fastening regime is adopted, the performance is excellent, especially under outside weathering conditions.

Some new applications of plastic composites may be worth mentioning. A glass-fibre automobile spring, in the single leaf design familiar in older vehicle suspensions, was adopted for the rear suspension system of the 1981 American Chevrolet Corvette car. This spring, weighing eight pounds and made of epoxy resin – glass-fibre, replaces a 41 pound ten-leaf component, Anon. (1980*c*).

Mention has been made of polyester liquid resins as a matrix material for glass reinforcement. Knight (1983*d*) discloses a new foaming system for these resins, which reduces the cost and improves the strength and stiffness of structures reinforced by glass fibre. The foamable resins can be applied by spray gun so that glass fibre reinforced components can be made by the old-fashioned hand or spray lay-up techniques. The product can have either a single or double skin of solid polyester resin with a foamed centre; alternatively one of the skins can be made of thin, vacuum-formed acrylic into which the system is sprayed.

Aromatic polymer composites based on polyether etherketone reinforced with carbon fibre are claimed by Cogswell & Hasler (1984) to be the first of a new generation of structural composite materials. The composite is prepared as a tape containing more than 50 % by volume of continuous fibre reinforcement, and the subsequent processing involves only heating and forming. This tape is available as a pre-preg or as sheet stock shaped by laminating at 380 °C under a pressure of 10 bar. This unusually high temperature is needed because of the high rigidity of the polyether etherketone, which produces composites with an impressive range of properties. For example, a uniaxial composite has a relative thickness of only 120 % of that of steel, with the same bending stiffness: put another way, the relative bending stiffness values for the same weight are 0.44 for steel as compared with 35 for the composite. This leads the way forward to a whole new generation of engineering materials.

Zweben (1984) reviews the principles of fibre-reinforced metal matrix composite systems. The potential of these materials for elevated temperature service is apparent but at present their viability is limited by fabrication difficulties and by some fundamental chemical stability problems which obviously have an impact upon the relevant economics. Early attempts to manufacture metal matrix composites using inorganic fibres were largely hampered by lack of suitable fibres having high temperature performance coupled with low cost. Of those currently used, alumina is now available as pure alpha-alumina fibre in the form of flexible, continuous multifilament yarn. This bonds very well to metal matrices, as for example, magnesium and aluminium, conferring interesting strength increases on these otherwise weak metals. Current application studies involve the use of the composite with magnesium for helicopter transmission housings, Dhingra (1981).

Recently a major advance, ARCO (1983), was claimed in producing a 450 mm diameter, 600 mm long, billet of Type 2124 aluminium alloy, reinforced with silicon carbide whiskers. Weighing 282 kg, it is described as the world's largest composite billet fabricated to date. From billets made of aluminium and of copper, titanium, magnesium and other alloys, it is possible to forge and flat-roll products found to achieve elastic moduli and strength up to twice those of the base alloys.

8

Materials performance in service

There is an ill-defined grey area of expectation between properties of materials as published by suppliers (or as determined in a producer's laboratory) and what is achieved in final product performance. Philosophically, the act of taking a raw material and converting it into the test piece from which a property parameter can be determined is in itself 'conversion to a product' and is arguably sufficient to alter the original properties. The performance of a material as a test piece may not necessarily truly reflect its performance when in a larger mass or in a different shape.

An example may be drawn from the plastics field. Suppose a thermoplastic test bar is injection moulded in order that an impact measurement can be made. The material will fill the mould surely enough, but in doing so will have some degree of molecular orientation impressed upon it by the flow process. Consequently the test bar will have a preferred tensile strength direction along its long axis; this must modify the measured impact strength as compared with a totally isotropic material which has, for example, been poured into the mould from a liquid melt. An injection-moulded article with a range of section thicknesses, stiffening webs and surface areas in different planes will have a different level of molecular ordering to that in a test bar. Under impact conditions the laboratory test result could not accurately be extrapolated to reflect a finished article. So even today, with an article like an injection-moulded plastic telephone subscriber set, the only accurate way of assessing impact strength is to drop it, in a number of well-controlled orientations, from a standard height onto a standard hard surface.

Another example is offered by cast iron. Suppose a cast test bar is made for assessing tensile strength. The low heat content of the melted iron, relative to that of the mould, will have a profound effect upon the rate of melt solidification and hence upon the extent of precipitation of carbon and other non-ferrous phase particulates in the bar. This structure may prove considerably more finely grained than would occur in a massive casting of iron, where the cooling rate could be considerably slower, with a consequent increase in the size of the precipitated phases. These foreseeable different internal structures in the materials suggest the possibility of different test results being obtained.

Such factors suggest a high degree of uncertainty in exploiting materials and should tell the designer that, in normal circumstances, he should never refine his design to be wholly dependent upon the apparent limits of a material's properties. A factor of safety should always be allowed; this is discussed in a little detail in Chapter 10. However, in some applications, such as in the aerospace industry, reduction of product weight is at such a premium that near-ultimate designs have to be accepted: this specialist area with its own philosophy of control does not fall within the purview of this book.

Normally, we should accept that there will be varying degrees of overdesign of a product in terms of performance of its material in mechanical, thermal or chemical ways; this will represent some uncertainty as to exactly what performance should be expected of the product itself. Having said that, we should now consider various service conditions and their effects upon materials of construction.

8.1   Metals

The intrinsic properties of metals alone do not govern their behaviour in service. Every material in practical use is subject to environmental influences around it; metal performance in service is dominated by the twin collossi of fatigue and corrosion, the former being a feature of the metal itself, while the latter reflects the influence of the surroundings. These concepts and some others are explored briefly in the following sections.

8.1.1   *General properties of metals in service*

As for other materials, metals may be subjected to mechanical traumas, either separately or in combination with active corrosive influences; due notice should have been taken of these when selecting the material, as discussed in Section 4.1. Metals as a generality have a number of strong points for which they are chosen, but they also have a variety of deficiencies. Strong points range from the attainable dimensional accuracy in parts, and their subsequent stability, to the smooth surface which can be achieved with die-cast zinc alloys. Strong points include good impact and fatigue properties coupled with the excellent machining characteristics of malleable iron. An avoidable deficiency in metals is the tendency to creep if high loads are maintained for long periods at elevated temperatures.

Metals are often selected for engineering applications on the bases of their true elasticity at relatively low strains, their useful surface hardness and their good performance in resisting the various forms of wear regime. A strong point may constitute a deficiency in alternative use areas. Thus whereas special magnetic alloys, and indeed ferritic malleable iron, can be used when magnetism must play its part in the operation of a product, this same magnetic characteristic may be a disadvantage if it causes metal parts to gather tramp iron dust which then wears out a moving mechanism. Similarly, although properly made zinc die-castings have high dimensional stability, some of them, as in the case of alloys containing copper, can suffer dimensional changes during ageing in which a few weeks' shrinkage after casting may be followed by subsequent expansion. This can be difficult to allow for, as tolerances essentially become variable during the product's early life.

8.1.2 Strength considerations

Metals are usually chosen by a designer because of their strength characteristics. Thus in the case of malleable iron, the yield strength, ductility, rigidity, fatigue strength and impact resistance can be exploited as long as limits of performance are well defined. Ductility is an oft-quoted parameter for metals, but it should be remembered that excessive ductility (in terms of a large measurable elongation under stress) is not usually necessary in a product, although it might be so in a test which defines the ultimate performance of the metal. The ductility should usually be sufficient that, at a high stress rate (impact), the material is sufficiently tough to elastically or plastically deform sufficiently to absorb the imposed energy. At the other end of the rate scale, ductility can play its part in achieving an equable spread of stress in ill-assembled products, where fasteners may have been unequally tightened to produce localised high stress points.

Fatigue strength is always a well-documented parameter for materials, which should be chosen to reflect known service requirements. However, that kind of random or cyclic vibration or movement which might cause fatigue failure could also lead to fretting of the metal, and this possibility should not be overlooked.

Metals are often selected for rigidity, based upon their modulus of elasticity in tension. However, in many designs, it is more important to consider specific stiffness in which the specific gravity of the material is allowed for. Sometimes the lower modulus material, aluminium, can be considered as a replacement for the steel which was first thought of.

Impact resistance of metals has been touched on before. Apart from ductility playing its part in an unnotched sample, it should be recognised that some metals are notch-sensitive, and the appropriate impact test data must be reviewed before a final choice is made.

8.1.3   *Corrosion effects*

Even the so-called clean air which humans breathe contains traces of materials potentially corrosive to metals in service. Studies of corrosion of electrical contacts have led to measurements, in industrial and rural atmospheres, of many pollutant gases, which show a wide range of concentrations. The gases nitric oxide, nitrogen dioxide, ammonia, hydrogen sulphide, sulphur dioxide, sulphur trioxide, hydrogen chloride and even free chlorine have been found in the atmosphere, as well as air-borne ionic dusts containing sulphates and chlorides which circulate freely around the world. All these materials can play their part in long term deterioration of metals.

Care must always be taken that corrosion tests effectively simulate the environment to be withstood by a product, but extra care must ensure that the acceleration factor used is sufficiently small for realistic deterioration mechanisms to take place. Excessively fast corrosion testing can lead to unnatural chemical reactions taking place which would never occur under natural long term conditions. Thus, for example, it is easy to expose a copper sheet to a corrosive atmosphere and to turn it green in a few hours, but this is not to imply that the test reproduces in any way the natural green patination which occurs on copper-clad building roofs over a number of years. In the two cases, the natural patina has a protective effect on the underlying metal, whereas the accelerated test, if carried out too fast, produces a different chemical corrosion product which tends to drop from the substrate metal and thus expose fresh areas to subsequent attack. In many cases also, when considering corrosion, care must be taken to choose the appropriate alloy of the generic metal under consideration. For example, malleable irons are more resistant to atmospheric corrosion than some carbon and alloy steels, especially in salt air, but the addition of a half-per-cent of copper increases the corrosion resistance still further.

Corrosion can occur on metals in many ways. General corrosion implies attack on a metal by reaction with its environment, producing a chemical compound which may or may not remain adherent to the substrate metal. This corrosion proceeds uniformly over the metal surface and can often lead to a predictable reduction in thickness with time. Intergranular corrosion is a selective form of attack in which the material at grain boundaries of metal is attacked so that individual crystals can fall away. Pitting corrosion is somewhat analogous but attack is localised, usually on impurity areas in the metal, and takes place over a very small area to produce what can be a relatively very deep hole. This may result in metal perforation in an otherwise structurally sound part. Crevice corrosion has a similar effect, in which a difference in chemical composition between a crevice in the metal and the outside world can stimulate attack. The crevice may be a sharply angled corner or recess, or perhaps the narrow space between two parts bolted together.

In demetallification, a corrosive agent can be selective and may remove just one metal from an alloy. Thus dezincification of brass (copper – zinc alloy) can occur so that a component retains essentially the same dimensions as the original, but has a porous structure and is without significant mechanical strength. The most general form of attack is galvanic corrosion, discussed in Chapter 9, particularly with reference to Table 9.1. This shows the electrochemical series for metals and, in a galvanic pair, indicates the metal which becomes the anode and is thus corroded away.

Even simple exposure of metals to ambient air can lead to galvanic corrosion, because the exposed surface will always bear a condensed layer of adsorbed water acting as an electrolyte, especially in the presence of soluble atmospheric gases such as carbon dioxide or sulphur dioxide, or in the presence of soluble dusts such as sodium chloride from the sea. Mention of the sea conjures up thoughts of corrosion of marine structures, often exhibited as pitting arising from relatively stagnant conditions, local deposits of foreign matter or the presence of heavy metal ions. Pitting ratings in sea water are well understood and have been tabulated (Waterman and Pye 1980).

Metal corrosion is often more mundane. For example, microorganisms can attack underground pipe systems, as shown by Smith (1981). The byproducts of microbial activity include acidic chemicals at high concentrations which can, in the long run, cause serious attack. Smith (1982) provides a resume of the very long established understanding of attack on metals by wood, this having first been successfully studied at the time of the First World War when military equipment was transported worldwide for possibly the first time in history, packed in seemingly innocuous wooden boxes. Wood causes corrosion of metals by direct contact and in confined spaces, usually through the emission of corrosive vapours such as acetic acid. Of the woods commonly available, it is interesting that oak is the most corrosive, while obeche is the least.

There are several ways of avoiding corrosion in metallic structures by making appropriate design decisions. These include avoiding stress corrosion, fretting corrosion and bimetallic corrosion, through proper design and choice of materials and environment, and by ensuring that contours and corners are rounded and unable to provide crevices for local attack. Turbulence and rotary motion in liquid environments can also lead to accelerated corrosion and should be avoided, or, alternatively, their corrosion potential catered for. Water should be prevented from collecting in restricted spaces such as in small areas on an outside surface, or through condensation in a hollow vessel. Finally, welding should be carried out sensibly and the use of non-identical metals in a structure should only be permitted with knowledge of the stray currents likely to occur between them in service (Smith 1980).

8.1.4   *Service temperature limits*

Finally in this brief chapter, mention must be made of the effect of service temperature on the strength of metals. It may not generally be realised that metals can lose a considerable proportion of their strength at temperatures well below the nominal melting point. This explains why a serious fire in a steel framed building must lead to ultimate demolition, unless the frames have been very well protected by thermal insulation. At temperatures around 500 °C, steel girders constituting a frame have a reduced elastic limit such that they are unable to support their own weight, leading to the tangled forest of buckled beams and spars so familiar in newspaper photographs. The temperature is not a magic number but varies with the material. Thus, for example, malleable iron is claimed to retain appreciable strength even at 425 °C and can support some loads at lower stress levels at 650 °C. The direct loss of strength through heat exposure and the consequent metallurgical changes in the material are functions of time. In situations where the material is not heated sufficiently for virtually instant destruction, as in the case of a severe fire in a steel framed building, there is still the question of creep as a function of stress, strain and temperature. Curves of these parameters for all materials are reasonably well documented. Creep is quite well withstood by some dispersion-hardened alloys, in which heat-precipitated and well dispersed phases of strengthening material are deposited in the base metal matrix.

In addition to fairly short term heat treatments, oxidation effects on metals can ultimately lead to total breakdown. Although aluminium is a good example of a metal bearing a coherent surface film of oxide which protects the underlying metal, this is not so with other materials such as iron, which when heated in air become covered with 'mill scale' oxide forming adherent coatings that look like strong metal, but which in fact spall off under moderate impact. The exposure of iron-based alloys to such oxidising conditions is not to be countenanced for engineering applications in service without additional surface protection.

8.2   Ceramics

Ceramics are inorganic, non-metallic solids which usefully complement composites, metals and polymers in the main classes of material used in engineering. Ceramics have a wide range of properties, many of which were sketched in Chapter 5 but which can briefly be summarised as refractoriness, hardness and stiffness. Special ceramics may be particularly good electrical insulators, or show piezoelectric behaviour in high-voltage alternating fields, or perhaps have interesting magnetic properties. However, these are not considered in the present context of engineering applications.

8.2.1 *Advantages and limitations*

The optimist will be happy with the thought that his design specifying ceramics will be rigid and stiff when necessary, and resistant to creep and high temperatures. Ceramics often operate at temperatures above those at which metals lose significant residual strength, and retain useful attributes including resistance to abrasive wear and to many chemical environments.

The strength of a ceramic in tension is generally not spectacular, the materials often acting in a brittle (otherwise termed 'non-ductile') manner. This is usually caused by the presence of crevices and faults on the surface or within the material, from which the fracture process can commence on applying a stress. This process is, of course, temperature-related because many ceramics will undergo permanent deformation at high temperatures and then are ductile. Morrell (1980) reports that defects from the manufacturing process lead to low values of the critical stress intensity factor which turns catastrophic propagation of defects into running cracks. The size of these strength-limiting defects is extremely small when compared with those in ductile metals, and difficulties in controlling their size and nature lead to a distribution of measured loads for fracture in any population of apparently identical test bars of ceramic, a situation not reflected in similar tests with metals. The consequences of brittle failure in test pieces can be interpreted in terms of the statistical (Weibull) distribution, which designers use to achieve predictable reliability levels in their designs. Baker (1980) explains how fine cracks in ceramic substrates can still be detected, even in samples which have passed hermetic seal testing. His method depends upon the condensation of a vapour on a surface viewed at a carefully chosen angle to the incident light.

Little need be said about the strength of ceramics in compression, except to point to the very familiar example of electrical insulator supports for the live rail for underground trains. Obviously, design must recognise the expected limits of creep for the various conditions of service which the material has to meet, and the material also must be chosen with a sufficient level of toughness that the amount of energy needed to form a new fracture surface exceeds that likely to be met under extremes of mechanical abuse. Although the compressive strength of a ceramic material may be several times as great as its tensile strength, as determined by a bending test, this value is not necessarily important because application of pure compressive stress is quite rare in mechanical engineering.

A further important mechanical feature of ceramics in service is their wear resistance. Some materials, including heavy metal carbides, have such good wear resistance and hardness that they are used in combination with a metal matrix binder to produce cermets for metal machining purposes. Cermets are often used in engineering because of the economically achievable wear resistance, coupled of course with a high degree of hardness normally measured in units of GN/m^2. Wear resistance as a parameter is much less well standardised than are hardness measurements; there seems to be an inherent variability in test results which, perhaps, reflects the different wear mechanisms that can take place.

Finally, in this brief consideration of mechanical properties in service, notice must be taken of fatigue. In metals, fatigue crack growth rate can be evaluated from well-known equations which show that the average increase in crack length per loading cycle is virtually independent of frequency. For at least some ceramics, different mechanisms apply in the cases of cyclic fatigue and of static fatigue, because the growth of cracks depends upon the frequency and the average stress intensity applied to the part.

Turning from generalities to an example: silicon nitride has a unique combination of properties with high strength and hardness, good wear resistance, stability to higher than 1800 °C in air, and with excellent resistance to thermal shock; all this with less than half the density of steel. These properties, in theory, make the material very promising for fabricating engine components, particularly in gas turbines. Unfortunately at present too much porosity and not enough dimensional accuracy in the components prevents this goal being achieved. The attempts, however, have led to improvements in silicon nitride ceramics, particularly through addition of fluxing agents which increase densification. The material continues to be increasingly interesting for other automotive engine uses, as well as for tooling, wire drawing dies, welding components and industrial wear surfaces. Jack (1984) discusses several potentially interesting uses for this material and its variant, which is a silicon − aluminium − oxygen − nitrogen system generically named sialon.

8.2.2   *General environmental effects*

Ceramics are well-balanced materials, chemically speaking, and usually have a low reactivity towards their environment. However, as Morrell (1980) points out, resistance to chemical attack is not to be considered as a well-defined intrinsic material parameter; it represents an observed qualitative performance which depends on various internal and external influences. It is very easy to identify the external causative agents because these involve tangible factors such as ambient gases or liquids, and easily measured parameters such as temperature and time. The ceramic material's internal constraints are not generally so well-known to the designer, who has not usually specified, and certainly not manufactured, the material as a chemical entity. He will accordingly be unaware of composition imbalances, the presence of crystalline or amorphous phases and the level and distribution of porosity which can influence chemical attack on the ceramic.

In general, and in contradistinction to metals, which frequently are dissolved by exposure to acids, ceramics may suffer very little weight change after chemical attack. However, the attack which does take place is most likely to occur at relatively weak grain-boundary positions, leading to a serious loss of strength but virtually undetectable loss of weight. It is mentally satisfying to know that ceramics may be corroded by liquid metals and by fused salts and slags, but perhaps not so obvious (yet more useful) to know that they can be attacked by aqueous media at room temperature. A well-known material in this context is hydrofluoric acid, which can attack silicate compounds in particular.

No exposition of environmental attack effects on ceramics is going to help the designer avoid all the problems. It should suffice, in the present text, that he is alerted to their possibility so that, as usual, enquiries can be directed towards the supplier or to a known and trusted expert adviser.

8.3 Polymers

Even in the relatively benign atmosphere of the home or office, 'normal' environmental factors can affect durability, performance and joy of ownership of an article made from plastic materials. The specific behaviour of plastics and rubbers in more demanding environments, as in aerospace, marine, chemical and industrial engineering applications, must first be determined and then offset by proper selection of materials.

Taking the use of polymers in civil engineering as an example, polymethyl methacrylate, polyvinyl chloride and polycarbonate are widely used for glazing, roofing and cladding of buildings, and polypropylene and polyethylene find uses in pipe-work and in manufacture of tanks. The materials can all be formulated to withstand outdoor conditions. Thermoplastic films find uses as impermeable membrane barriers to capillary water movement, while both woven and non-woven geotextile fabrics based on polypropylene (alone or with polyvinylchloride) are used as permeable membranes for ground stabilisation and containment. Such materials must have high stability to water. Fabrics are also used as exterior walls for free standing buildings, kept in place by internal air pressure, and often based on woven glass fibre cloth coated with polytetrafluoroethylene which not only resists weathering but which is self-cleaning in rain.

Other plastics used as surface coatings delay corrosion of metal structures, are the main ingredient of adhesives, and as additives to polymer concretes. When one additionally considers the use of acrylamide as a chemical grout for soil and the use of PTFE in expansion bearings and resilient mountings, it must be evident that a very wide range of plastics is used in this one sphere of civil engineering, with all the implicit consequences of exposure to a wide range of environmental conditions.

In service, plastics may be exposed to temperatures ranging from those attained on the surface of a space re-entry vehicle down to those sub-zero temperatures imposed by space operations or by cryogenic systems. Ultraviolet light and other electromagnetic radiation is always present, depending upon the latitude and altitude of service, and can have significant influence even in the murky temperate zones on earth. Contact with water can vary from permanent immersion in sea water to wet – dry cycles in the tropics. These considerations must naturally lead the designer to consider the main types of environment which might affect plastics and elastomers in service. Some of these are defined in the next few paragraphs.

8.3.1   *Chemical environments*

The term 'chemical environment' embraces not only the effects of the most aggressive environments in service, but reflects also those atmospheric materials present during natural weathering. Even moisture, humid conditions of atmospheric humidity or of total immersion can act as an aggressive service risk.

Some plastics when under mechanical stress are sensitive to cracking during or after contact with certain organic liquids which attack crystal boundaries in the material, producing areas of localised high stress and leading ultimately to overall weakness. The result is usually that the material will eventually fail through brittle fracture, which would not have occurred under similar conditions of stress in the absence of the liquid. This effect, of course, is more obvious when vibration, cyclic stress or shock are normal ingredients of the environment during service.

8.3.2   *Mechanical environments*

Any material in mechanical service is likely to sustain, possibly simultaneously, periods of long-term sustained stress, varying cyclic stress which can induce fatigue failure, acceleration and impact forces. These mechanical traumas lead to creep effects which are exacerbated by exposure to elevated temperatures. Excessive mechanical force produces dimensional changes and possibly ultimate malfunction or demise of the product.

A material or a component may be said to have failed when:
1. The strength and other physical properties fall to an unacceptably low value, from either reversible or irreversible causes.

2. Through material changes, it is unable to resist additionally imposed mechanical stress, or physical or chemical environments without irreversible further degradation.
3. The material or component cannot properly perform the functions for which it was designed, and which it could perform when new.

8.3.3 *Biological effects*

Engineers very often ignore or underestimate the effects of biological agents on service performance of materials. A recent example, not concerned with construction engineering, is the realisation that mould growth can occur in fuel tanks of jet aircraft, giving rise to moisture retention and the possibility that pieces will break off and clog fuel lines. More obvious attacks are from staining microorganisms resident on stone surfaces of buildings, and insects boring into plastic materials. For example, even in parts of temperate Europe there is a strong possibility that termites will irreversibly damage unprotected polyethylene insulation in buried cables, while attack on such cables by the teeth of gophers is well-known in certain parts of the USA.

Biological degradation can result from the depredations of many vectors, from bacteria through moulds and fungi to insects and mammals. Because organisms often see a polymer as food, materials or additives which offer good resistance or even immunity to biological attack will incorporate repellents such as compounds of sulphur or of halides (chlorine or fluorine), or any chemical groups upon which enzymes or microorganisms can act will be absent from their molecules.

Although certain types of polymer are known to be immune to microbiological attack, while others are chemically predisposed to be eaten or grown upon, nevertheless selection of materials recognising these criteria alone would be far too restrictive to a designer when choosing a plastic material for product use, except in very specialist applications. Formulations of polymers often include fillers, stabilisers and plasticisers. Fillers, which might be as cheap and as 'low-technology' as sawdust or starch, might well be attacked by microorganisms. More important, however, is possible attack on the plasticiser generally used in flexible polyvinyl chloride products. Plasticisers are usually long-chain organic acids or esters, well able to support the growth of microorganisms, so that attack may occur on this material rather than on the polymer itself. The result is that the plasticiser is consumed and the host plastic either disintegrates or becomes increasingly brittle and less ductile. Some internally plasticised thermoplastic resins, on the other hand, are resistant to microbiological attack because a second readily separable phase of material is not available to the predator.

The lower hardness of plastics compared with many other materials makes them subject to disruptive attack by insects or rodents. The object of such attack may be to provide bedding material, to remove the offending object from a preferred burrow, or even merely as a convenient means of sharpening teeth! Lovelock and Gilbert (1975) cover the microbiological degradation aspect of materials deterioration by discussing the organisms needed for biodeterioration testing, although this will normally not concern the designer.

8.3.4   *Weathering effects*

Combinations of solar radiation, humidity, temperature, contaminating atmospheric gases, precipitation of solid or liquid water, and erosion by wind-borne particles may be experienced by materials during natural weathering. Apart from the manifold possibilities of chemical change in even a static system, the rate of degradation may be influenced by dynamic factors such as thermal cycling and wind-speed effects. Moakes & Norman (1977) present information on typical atmospheric parameters which can be met in four representative world climates, referred to in Table 8.1.

The service deterioration of compounded plastics and rubbers depends on the sensitivity of the basic polymer to chemical agents in the environment, on the particular technology used in fabrication, or the presence of additives perhaps included for other purposes, and even on the geometry of the article exposed. Accordingly, apart from climatic factors, other environmental mechanisms can readily operate on these organic based materials.

It should be remembered that most plastic and rubber synthesis and compounding activity takes place in developed countries, for markets primarily in similar climatic regions. Materials for service in temperatures far removed from the 'room temperature=23 °C' concept are usually made as lower

Table 8.1. *Some world climatic conditions*

Arctic	Temperate	Desert	Tropical
Exposed arctic temperatures as low as −40 °C to −70 °C	Climate is intermediate between tropical and arctic	Air temperatures range from +60 °C to −10 °C Surfaces exposed to direct rays of the sun may reach 75 °C Maximum relative humidity ranges from 5% to 10% inland	Air temperatures range from 20 °C to 52 °C Surfaces exposed to direct rays of the sun may reach 70 °C Relative humidity in excess of 90% at night

volume specials. These special formulations take account of different rates of reaction of materials under the prevailing local environments. Temperatures above the temperate region norm, in the presence of oxygen, can progressively initiate chain-scission in the polymer, and promote oxidative reactions which produce smaller molecules and hence a loss of characteristic properties, especially viscosity and strength. High temperatures may also accelerate photo-degradation reactions, in which ultra-violet energy at wavelengths between 290 and 400 nanometres (about 6% of the total solar radiation reaching the Earth's surface) triggers chain-scission, cross-linking and oxidation reactions in plastics and rubbers. Some polymeric materials absorb radiation uniformly over the whole ultra-violet spectral region, and then no clear-cut relationship between wavelengths and consequent damage is apparent. Other materials absorb ultra-violet radiation only at specific wavelengths, and may be damaged to a proportionately greater extent under some exposure conditions.

Hydrolytic and chemical attack also tend to convert a polymer into smaller molecules, or even operate by physical effects, for example the direct loss of additives such as organic plasticisers. Chemical degradation results from the

Table 8.2. *Environmental factors affecting plastics and rubbers*

Parameter	Manifestation and effects
Radiation	Solar, nuclear, thermal
Temperature	Elevated, depressed, cyclic around the 'norm'
Physico chemical factors	Chemical attack, physical changes such as plasticiser bleed
Organic solvents	Vapour absorption, dissolution, stress corrosion cracking
Stress factors	Sustained stress, cyclic stress, compression set (in rubbers) under continuous loading
Biological factors	Microorganisms, fungi, bacteria, animals, insects can destroy materials or change their properties
Wind	Air-borne particulate erosion from dust or water precipitation
Normal air constituents	Oxygen, ozone, carbon dioxide, nitrogen oxides
Air contaminants	Gases: sulphur oxides, halogen compounds Mists: aerosols, salt, alkalies Particulates: sand, dust, grease
Combined action of wind and water	Surface damage
Water	Solid (snow, ice), liquid (rain, condensation, standing water), vapour (relative humidity). Rain, hail, sleet or snow may have physical effects
Freeze – thaw	Thermal expansion
Use factors	Normal wear and tear, abuse during installation, abuse by user application outside the designed use conditions

attack on polymers by ozone, sulphur dioxide and nitrogen oxides, which occur naturally in the atmosphere but more probably arise in greater amounts as byproducts of man's activities.

The failure mechanism of a material depends on the chemical, physical or mechanical properties, and the processing technology, as well as on the environment to which it is subjected.

Table 8.2 summarises the exposure factors which can lead to reversible or irreversible changes in polymeric products. Specifically, natural weathering factors are illustrated in Table 8.3.

Although absolute confidence in the reliability of a product can only be assured by long-term natural weathering and service trials, such experiments can rarely be undertaken in this time and cost- conscious commercial world. Before such information becomes available, selection of materials and design data decisions have to be based upon laboratory tests, coupled with experience from previous design decisions. There is still controversy over the validity of accelerated laboratory tests for predicting performance in service. Accelerated tests induce failure at a rate many times faster than normally occurs in service, producing information in a relatively short time. However, results will be misleading if the correction factor for the accelerated test rate cannot reliably be extrapolated back to the slower rate situation which occurs in nature. There is always the danger that to achieve a reasonable acceleration factor the test conditions may be so enhanced that new reactions may occur in the test piece which would never occur in normal usage.

A byproduct of both laboratory and natural testing is the ranking of materials in order of usefulness for each parameter of consequence in the design of a product. Unfortunately, the relative ranking order of a set of materials depends radically upon the particular parameter of interest; never does one find that a given material ranks, for example, third in a list of ten for every one of the

Table 8.3. *Some effects of exposing polymers to solar radiation, humidity and temperature changes*

Solar Radiation	Humidity	Temperature
Stresses from cross-linking reactions	Plasticisation Moisture gradients Swelling gradients Leaching of unreacted material or additives Absorption, desorption Swelling fatigue stresses	Thermal gradients Thermal stresses Thermal fatigue stresses Evaporation of unreacted materials Loss of volatile additives

parameters of interest. Materials are not like that; there are no 'best' or 'worst' materials; each has its own mix of specialities to offer, and the designer's challenge is to recognise this and to choose the property combination best suited to his needs.

Some of those properties of polymers and elastomers often evaluated by laboratory tests to simulate natural weathering are listed in Table 8.4.

The usual environmental evaluation regime exposes a test piece to the required conditions and observes the time elapsed for the first significant change in properties to occur, subsequently following the rate of deterioration.

Table 8.4. *Properties used as indicators to assess likely resistance to natural weathering*

Strength
        Tensile properties (particularly elongation at break)
        Flexural properties
        Impact strength
        Tear strength

Dimensions and mass
        Length, width and thickness
        Warping
        Delamination
        Change of mass

Appearance and surface properties
        Colour
        Gloss
        Haze (for transparent materials)
        Blooming and exudation
        Crazing and cracking
        Biological attack

Electrical properties
        Loss tangent
        Dielectric breakdown
        Surface resistivity
        Volume resistivity
        Permittivity
        Arc resistance

Bulk properties
        Thermal conductivity
        Water absorption
        Water transmission
        Porosity
        Bond integrity (at interfaces in composites)

An example of this in Table 8.5 shows how the mechanical properties of an epoxy resin − glass-fibre composite were affected by six years of unstressed outdoor exposure in Ohio, USA. The results confirm that tensile and flexural strengths are substantially retained, but compressive strength is significantly reduced. Nonetheless these changes might confirm that, used as an outside decorative panel of a building, an epoxy-resin − glass-fibre composite could be suitable, at least in the short term. The effects of weathering on mechanical properties are of major importance in most applications; tensile properties have long been used for quality control purposes in plastics, and elongation at break (a measurement taken at the same time) can indicate the transition from a ductile to a brittle condition. In many weathering situations, however, breaking stress may remain relatively unchanged or even apparently be increased, although there would usually be marked deterioration in other physical properties such as flexural or impact strengths. Of more practical significance perhaps for evaluating degradation by laboratory tests, would be the measurement of yield strength and yield strain. Flexural strength is markedly affected by the state of the specimen's surface, and is therefore a useful parameter by which to determine the onset of degradation, particularly when the exposed surface is put under tension. Sometimes measurements in which applied stress is held constant for different periods can provide information on the likely service life of a component which must sustain well defined forces. Table 8.6 shows how the expected service life of a laminate increases as the stress upon it is reduced.

Another property of practical importance is impact resistance, also influenced by surface degradation. In assessing this property, tests on notched specimens are considered to be reasonably representative of those residual stresses which can be found in moulded plastic components.

Electrical measurements may be necessary for certain applications in which weathering may lead to a decrease in electrical resistance, particularly if moisture is absorbed, as is the case with polyamides. Furthermore, electrical

Table 8.5. *Retention of mechanical properties of epoxy resin − glass-fibre composites after six years outdoor exposure*

| Material | Percentage property retention | | | |
	Tensile strength	Flexural strength	Compressive strength	Interlaminar shear strength
1	99	86	38	81
2	96	90	57	87
3	84	95	45	91

measurements such as surface tracking or arc resistance may be called for when, for example, a material has had incorporated that type of flame-retardant additive which diffuses to the surface and which might modify surface electrical properties before it diffuses out and is lost for ever.

Attention has not, so far, been focused on the performance of elastomers, in which the most significant indicator of deterioration tends to be the measurement of ultimate strain and modulus. Table 8.7 illustrates values available in the literature for a range of elastomers, showing how some of them can be so seriously affected by exposure as to be unsuitable for service in extreme environments.

Summarising, it can be stated that, degradation processes in polymers and elastomers are even now not totally explainable although there is considerable understanding of certain details of reactions of these materials with mixed environments such as combined temperature and humidity, or with various forms of radiation. As natural conditions cannot be reproduced perfectly in the laboratory, there remains some small doubt about the validity of accelerated test results obtained by this route, being expected as they are to extrapolate what would be seen over years of normal service. Accordingly, it is still important for the designer to consider the environmental situation in which his product will be used, at a very early stage in the design process when materials are being selected. The designer has the options of working on the basis of probability of failure under the totality of the service environments likely to be experienced by the material, or, alternatively (and more usually), of overdesigning the product in his chosen material to ensure adequate long-service performance. It follows that decisions must be made as to whether to believe a worst possible performance prediction from accelerated laboratory tests, or whether to accept pre-existing data from natural or service trials, which may not necessarily reflect

Table 8.6. *Time to failure of an epoxy resin — glass-cloth laminate in tension at 23 °C*

Applied stress (MPa)	Time to failure (hours)
470	0.01
297	1.00
290	10.00
283	100.00
269	1000.00
259	10000.00
257	20000.00

Table 8.7. *Percentage change of tensile strength and elongation at break of rubbers after 15 years ageing in natural enviroments*

Rubber type	Natural rubber	Styrene-butadiene	Butyl	Chloro-prene	Nitrile	Acrylate rubber	Chloro-sulphonated polyethylene	Polysulphide	Silicone
Ageing regime									
Temperature ageing									
Tensile strength	−20	No significant change	−15	10	20	No significant change	No significant change	5	35
Elongation at break	−35	−25	−20	−10	No significant change	No significant change	No significant change	No significant change	No significant change
Tropical ageing									
Tensile strength	−20	−5	No significant change	10	15	No significant change	5	5	35
Elongation at break	−45	−40	−20	−10	−20	No significant change	10	No significant change	No significant change
Desert ageing									
Tensile strength	−50	−5	No significant change	10	30	No significant change	5	5	35
Elongation at break	−60	−40	−20	−10	−20	No significant change	−5	5	No significant change

those conditions to which the product of present interest may be exposed. This discourse cannot provide definitive answers for the designer, but is intended to convince him of the complexity of the problem and to make him seek guidance from experts in this field.

8.4 Composites

8.4.1 *Strengths and weaknesses of composite construction*

On the whole, fibre-reinforced plastics, rubbers, metals and ceramics are relatively new classes of materials, still accelerating towards a peak of perfection, which mostly do not have a very long history in terms of years. In properly designed composites based upon them, each component contributes its own desirable properties to yield a material in which, hopefully, required properties are maximised and reasonable compromises have reduced undesirable features. Composites can be designed to retain adequate strength in both mechanical and chemical environments. An example from chemical process plant is the use of glass-fibre-reinforced plastic for ducting airflows and for fabricating some types of storage vessel. The same material demonstrates its versatility in making very large shapes, such as boat hulls which, relative to conventional techniques, are cheap to manufacture, easy to repair and perform well in service.

In recent years, the many advantages seen in those hybrid composites which combine two or more reinforcement fibres in a simple laminate have expanded their use from the aerospace and defence industries into oil drilling, energy industries, transport, storage, fibre optics and electrical applications. Sprecher (1984) provides limited information on the main types of reinforcing fibre, including carbon, polyaramid, E-glass and magnesium oxide-alumina-silicate, which can be combined to tailor ever more accurately the mechanical and other properties of a required composite structure. Less exotic is the recent announcement, Anon. (1985), of a composite automotive tow-bar claimed to be as strong as a conventional steel one, but weighing only one-third as much. This was developed for the US Army and is based upon a carbon-reinforced epoxy resin tube claimed to withstand loads in excess of 582 kN in tension or in compression. The tube was wound on a mandrel made of resin-impregnated woven polyaramid fibre, retained to play its part in the structure.

Table 8.8 illustrates some of the characteristics of composites to assist the designer in formulating questions for his expert advisers. He should remember how unlikely it is that a standard part can be purchased off-the-shelf to take its place in a composite product. At best, one might purchase tubes, but designs will often require that the composite component be specially fabricated by a supplier for the purpose. His advice should be heard, checked and, if satisfactory, accepted as the final arbiter of the inevitable compromises in design.

8.4.2   *The matrix-reinforcing bond*

The key feature of composite mechanical performance is transfer of stress from the relatively deflectable matrix to the fibrous or particulate reinforcement which constitutes the relatively strong component. Transfer of stress can only occur if there is an adequate bond at the interface between them. The matrix is selected to resist thermal and chemical or other environments, to

Table 8.8 *Some advantages and limitations of composites*

Advantages	Limitations
Complex shapes are easy to make with some forms of composite, e.g. glass-fibre-resin spray or lay-up	Fibre-reinforced plastics for injection or compression moulding may show unacceptable fibre breakage or orientation arising from the production process
High strength/weight and stiffness-weight can be achieved if high strength, well-bonded fibres are used	These materials tend to be expensive. Glass-fibre reinforcement lack of stiffness can partly be compensated by configuration or by use of stiffeners
Fabrication costs can often be low	Some raw materials such as carbon fibres are relatively expensive. Products may be variable if large scale labour-intensive methods are used in fabrication
Thermal expansion coefficients can be reduced to levels comparable with those metals, facilitating reliable service in composite constructions	Designs may have to accommodate anisotropic properties. Shrinkage stresses during curing of some resins can lead to sheet buckling
Fastening may often be achieved by the use of adhesives	Mechanical fastening can lead to stress raising at drilled holes, delamination and water sorption
Porous materials can be made with good thermal insulation values	
Considerable flexibility available in maximising desirable properties	
Chemical resistance can be high on unbroken surfaces	Even with metal or ceramic composites, the matrix material performance in service may offer a limitation to use
Composites tend to absorb vibration and impact well	Less than perfect control of matrix-reinforcement interfacial bonding can lead to variable and unpredictable weakness becoming manifest in service
Electrical conductivity can be achieved by incorporating adequate loadings of of metal whiskers, carbon fibres or carbon particles	Unless heavily loaded, intermittent particle contact through flexure or thermal cycling can lead to erratic performance although it is usually electrically good enough to dissipate static charges
Flammability can be controlled by incorporation of combustion or ignition retardants	Some fibrous reinforcement can produce a 'candlewick' effects in combustion and increase the fire risk

bond the fibres together, and to protect the fibres and the fibre matrix bond. Metal matrices may make a marked contribution to the total strength of the composites but, broadly speaking, the purpose of a composite is to work on the known and predictable properties of the reinforcing fibre bundle or particulate mass, the matrix properties being an extra, though calculated, bonus.

A particularly appealing example of transfer of stress lies in the case of elastomers reinforced by short fibres. Stone & Campbell (1984) demonstrate the design concept in an all-rubber dock fender illustrating, by a novel mathematical treatment of their data, the way in which treated cellulose fibres incorporated in rubber are able to transform the composite performance.

8.4.3   *Effect of service conditions*

Correctly fabricated composites in service are distinguishable by a fracture toughness which is high compared with the matrix material alone, this being achieved by transfer of stress through a good interfacial bond to the reinforcing material. However, reliable achievement of this in practice, does depend to some degree on the workmanship and the quality of the manufacturing process. For example, air bubbles at the interface (even if deliberately included there to produce a thermally insulating foam) present a rise in debonding and general material weakness, particularly in bending modes. On the other hand, the lay-up method for a composite can have an improving effect, particularly in respect of impact performance. An example of this is the recent manufacture of Royal Air Force radome covers by a stacking process partially developed by the Royal Aircraft Establishment at Farnborough. This process involves assembling stacks of interleaved layers of polymer, bonding film and reinforcing glass cloth. On compression moulding, the polymer flows to form a homogeneous matrix with excellent wetting of the glass, the fibres of which have been aligned according to a prearranged geometry. This technique, using high grade thermoplastic polymers, can yield a significant improvement in impact resistance, the reinforcing invariably providing better performance than the virgin polymers.

The effects of temperature are reasonably predictable. Usually an organic matrix composite will have a maximum thermal expansion coefficient equivalent to the virgin polymer, this, however, being reduced by the ameliorating influence of the (inorganic) reinforcing fibres or particles. By proper control of loadings a close match can be achieved with thermal expansion values of adjacent metals, thus ensuring that bonds between composites and metals can be maintained under all reasonable service conditions. Such close matching of thermal expansion is important if liquid water or its vapour are to be excluded from composite hermetic packages.

Chemical and biological attack are reasonably predictable if one works on the basis of a perfectly manufactured piece part, in which the whole of the outer surface is an essentially virgin, even though very thin, matrix material. The performance of this material will dominate resistance to attacking chemicals, at least until the surface skin is corroded away and the particulate or fibrous reinforcement becomes exposed. At this stage it is quite possible for significant and rapid attack to occur at the reinforcement-matrix interface which often is exacerbated by the prior application to the reinforcement of chemically distinct primers with their own responses to chemicals. In general, once the reinforcing is significantly exposed, the resistance of the product against any chemical environment is seriously under question.

Similar considerations to chemical attack occur in the case of erosion, where the matrix surface coating is gradually eroded away, showing a fibrous structure open to attack from the environment.

Nevertheless, despite all that has been said of disadvantages, composites have a vital role to play in modern society, and the designer should not be deflected from specifying or using them. Instead, he should be aware of potential problems so as better to guide his discussions with expert advisers.

8.5   Flammability

No excuse is offered for including a section on flammability of materials in this book. Despite sterling efforts over many years, resulting in very well defined and reproducible laboratory work relating to the flammability of materials in all their forms, fire engineering still remains an inexact science. This is demonstrated by the existence of an enormous number of standardised flammability test methods which are in use today.

That there should be so many is not surprising, when it is considered that the process of material combustion depends upon many interrelated factors. Tangible products are made of tangible materials, which may be of a single or of several kinds, offering between them a wide variation in response to fire, broadened by a wide variation in geometry. Taking this in conjunction with the extensive range of service conditions that has been outlined in earlier chapters, one can see why the product designer may well be bemused when facing the problem of choosing specific materials for construction, especially bearing in mind the impact of current public health and safety legislation and the general social climate of personal and environmental safety which obtains at present.

The term 'flammability' is all about materials: materials in special shapes combined with others in sub-assemblies, which in turn may be assembled into products. Although, in principle, this takes us from an article the size of a small plastic gear wheel up to that of a large building, in fact the coverage of this chapter is confined to manufactured products which may be as large as a small room; it excludes, for present purposes, flammability control for large structures such as vehicles or buildings. Detailed advice is offered to those who must be concerned with possible fire hazards arising from products under their control but who, despite having to specify materials and configurations in designs, have not had any real training in the disciplines of flammability control. A reasonable understanding of the content of this chapter should at least help the reader to hold a sensible dialogue with an expert consultant to elicit the information he really requires.

Fires in small products are very often caused by electrical malfunction under operating or abusive conditions; this premise is implicit as the basis of the advice given. However, once started, a fire is a fire, so the recommendations apply equally to the possibility of preserving any product from ignition sources other than electrical ones.

Information is presented on selection of materials, the uses and abuses of flammability tests, and meaningful interpretation of their results. Before entering into a detailed exposition, it may be useful to consider the chain of activities in which a designer should be involved when considering flammability control for his product.

Firstly he should check whether customers, or current legislation in his market area, already specify what must be done. If this is the case, then it is very important indeed that the designer makes sure that the requirements are relevant, reasonable from both technical and economic viewpoints, and are also understandable. If problems are foreseen then, obviously, iterative discussions with the customer are important before going any further.

Next, if the designer has complete freedom to select materials and also has configurational control of the product, then the various suggestions in the following section on equipment design should be considered.

Then, when selecting materials, note should be taken of the pitfalls which might be met even with modern ones; this too is covered in Section 8.5.3 of this chapter. Materials which incorporate additives to inhibit the likelihood of ignition of a heat source (hereinafter termed 'delayed ignition') should be chosen, bearing in mind the effect that the presence of these additives may have on some of the desirable properties. For example, excessive incorporation of hydrated alumina as a fire retardent may make a plastic unmouldable; addition of halogenated flame retarding additive may spoil electrical properties such as loss angle. At this stage information should be available on the response of candidate materials to flammability tests relevant to the use situation. Such test information should be available before the design is frozen; manufacturers' test data (as long as the method is properly specified) is usually adequate at this stage.

Next, when reviewing the configuration of the product, consider how and where a fire could start and how it may then continue to travel or be stopped by the geometrical factors of the design. Identify also those test methods which might be used to validate the flammability hazard aspects of the product, and consider the limitations of the test methods; these are discussed later in this chapter. An irrelevant test method might at best yield misleading results, but at worst could be positively dangerous. It may be best to simulate a real-life hazard-based test for a specific situation, although care, judgement and experience are necessary to understand the limitations of such 'home-grown' test situations.

Then, as a further stage, it will probably be necessary to run a qualification test on the finished article, as required by the customer or by the designer's employer's policy. In the case of very expensive products, the qualification test may be carried out on a completely or partially finished mock-up, equivalent to the original from a material flammability point of view, although perhaps not equivalent in other ways such as electrical performance. Such tests may need to be made by an experienced consultant laboratory.

Lastly, if, after reading this chapter, there still remain problems which are too challenging for local resolution, then recourse should be made to an expert in the field.

8.5.1   *Customers' requirements*

Many professional customers, particularly the Military, Government Departments and Public Corporations, will already have a philosophy and specific requirements governing the flammability of equipment which they buy. As a matter of commercial sense, the designer's first impulse would be to ensure that these requirements are completely met. But think a little; these requirements might not necessarily be relevant to a new design concept of a new product. For example, the customers' requirements, geared to the flammability testing of a colour TV set, may be unthinkingly applied to a data modem marketed by the same organisation. Apart from the similarity of two boxes containing electrical parts, there is no way in which fire potential or fire response would be similar in the two cases.

The designer or his equivalent should consider his own viewpoint and be prepared to discuss with the customer incompatible requirements might need to be changed. For any changes to be acceptable, of course, they must be technically and economically valid, and all parties concerned must be satisfied that the requirements can be met and that the proper test methods have been stipulated.

Discussions with the customer should embrace overall requirements for flammability control in the product, the tests which should be used for materials, components and the complete assembly, and the pass-fail criteria by which the results should be assessed. The customer should be required to supply full information on the likely service life and maximum service-temperature requirement, as these might affect the choice of additives included in materials to delay ignition or to suppress smoke during burning. For example, regular in-service exposure to cleaning fluids might severely restrict the range of fire-retarding additives which could be used. Other aspects to be considered include provision of electrical failure control of high voltage malfunction; such malfunction can in some circumstances exacerbate fire hazards, as discussed later. The possibility and effects of toxic fume or smoke emission during worst-case prefire and fire conditions should be considered, as this may have some effect on materials selection and additives to be specified. Furthermore, corrosive gases could cause electrical arcing in areas remote from the first fire, leading to a separate secondary ignition.

8.5.2 *Product design*

Within the context of fire hazards in electrical or electronic products, nearly always the start of a fire will be internal and caused by some electrical malfunction. One cause would be through overheated electrical components suffering current or voltage overloads, which might be initiated by a fault in the component itself or in associated components. Another cause would be arcing or tracking across surfaces from broken conductor joints, loose contacts, deposition of particles or electrolytes between points of high voltage, or shorting of conductors across a high current drain.

Confirmation from the customer should be sought as to whether the designer's planned control of product flammability is preempted by market place or ultimate customer standards and requirements. These could include conformance with national standards, with international standards or with local practices of trade and industry, which themselves may stem from recommendations by a regulatory professional body operating within that market area.

Specification of totally non-burning or of extremely fire-resistant materials would be too restrictive for any but special constructions. It has to be accepted that many modern materials of construction are organic, such as wood and paper, and thus are always likely to support fire under the right circumstances. This is especially true for plastic materials which include plastic mouldings, wire and cable insulation, printed board laminates, structural plastic foams and others. The possibility must be faced that the material chosen may show a rapid rate of burning if the worst happens, or that it may release burning droplets which remain alight and so ignite more flammable substances below. This brings in a first tenet of product design, which is to be careful about grouping flammable materials in the lower part of the product, and their relation to those higher up.

Materials of construction, once mechanical, electrical or other performance requirements have been considered, should then be chosen from those with delayed ignition characteristics, these in many cases being readily available in the market place. The finer detail of materials selection will, however, have to be deferred until Section 8.5.3 of this chapter. The fact that a supplier claims that his materials offer a reduced chance of ignition from a standard heat source, or a slower than normal burning rate once ignited, will normally be vitiated by the possibility that, in a given product, materials sections may be much thicker or thinner than is used in the standard test bar required for the cited flammability test. For example, the well-known UL 94 flammability ratings are often quoted for different material section thicknesses, yet the test method is very specific about the unique dimensions of a valid test sample. Composites including, for example, glass-fibre reinforcement may transform a plastic material into a candle, in which the wicking effect of the fibres dominates the burning behaviour and ensures that combustion, once started, is total.

Still on the theme of an electrically based product, individual electronic components such as resistors, capacitors or printed wiring boards are usually supplied from elsewhere, and their individual flammability performance would, in a perfect world, be defined by the supplier on the basis of a given test. However, for articles of this type, there is no single standard acceptable test which provides unique and useful answers. It is often up to the product designer to look at the range of tests which is available, and to specify which of them can be used to provide results considered acceptable. Section 8.5.4 of this chapter discusses such tests in some detail.

Fire may arise in products from overloaded and glowing resistors, from small flames originating perhaps from faulty capacitors and relays, or even from arcs generated from high-tension or high-current power supplies by tracking across surfaces or through air gaps. The designer should lay out his components with these considerations in mind. Obviously, if a plastic material of known high and uncontrollable flammability is to be used for mechanical or other reasons, then one would not normally site it adjacent to a component likely to become overheated from a modest electrical fault condition. The designer must arrange that a component, which may self-ignite through electrical overloading, can glow or perhaps even burn out without the risk of overheating adjacent surfaces. He must also consider whether an electrical component not yet energised may nevertheless be ignited like any other combustible object by heat flux from an already burning adjacent component and may itself then contribute as fuel to an extended fire. It must be recognised that any component catching fire in an assembly may ignite several adjacent components and thus initiate a chain reaction causing destruction of the complete equipment and later on other combustibles in the same room.

In an ideal world, good layout design offers the ideal situation in which heat flux emitted from any burning component would be insufficient to cause any adjacent surface to become hot enough to start burning in its own right. This situation is illustrated by Figure 8.1. In principle, it should be possible to develop a computer-aided design system in which the total heat content and output rate of a burning mass of material could relate to the heat flux arising from it at different distances, and relate in turn to the absorption of heat by an adjacent material of known surface area, ease of ignition, specific heat and thermal conductivity, so that unacceptable surface temperature rises could be avoided. It would be helpful to point the reader to such a system, but such does not seem to exist at present, although available ISO ignitability test data does offer some help.

Obviously, in an equipment fire, a component or other material must start burning at one point and then may suffer flame spread over its whole surface.

Fig. 8.1. Factors influencing ignition. (Copyright: 1984 Standard Telephones and Cables plc.)

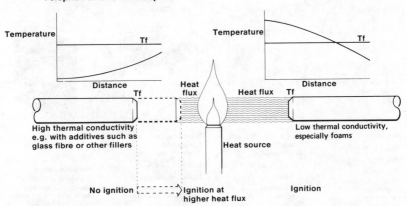

Tf is the decomposition temperature of the organic sample to yield volatile flammable gases. Only when the surface reaches this temperature, arising from a heat flux, can the test piece sustain combustion. If the sample receives a higher heat flux, surface temperature increases and flame travels along the sample.

An early British Post Office concept of classes of flammable material is illustrated in Figure 8.2, indicating that highly flammable materials must be used in the design only with discretion, but that modified materials with a sufficiently high limiting oxygen index (LOI see later), should be capable of slowing ignition at any one point and then delaying its subsequent destructive travel. Fires grow slowly at first but a pseudo-exponential growth can occur, leading to a 'flash-over' situation.

Close attention should be directed towards adequate circuit breaking control of electrical equipment. Any one component, malfunctioning to the extent of breaking down and emitting small flames or becoming red-hot, may not itself draw sufficient extra current in that condition to be capable of blowing any single fuse which often, alone, protects the whole equipment. Localised additional circuit breaking should be adopted in high-risk areas, and is even more important in the worst case where high voltages are in operation, since fields of the order of 500 volts per centimetre, spread across a gap in which there is a flame, and assuming an adequate oxygen supply, can effectively raise the base flame temperature and so accelerate ignition of any material in it. Because of the higher flame temperatures, there is little hope of staving off ignition of any organic material in this flame, however well treated with additives the material has been. Under such circumstances, all the painstaking effort expended selecting and using materials with delayed ignition additives is set at nought. Indeed, in service conditions where high temperatures (but not

Fig. 8.2. Flammability classes of materials. (Copyright: 1984 Standard Telephones and Cables plc.)

Class 1	Class 2	Class 3	Class 4
No burning	Burns upwards, not horizontally or downwards	Burns upwards and horizontally, but not downwards	Burns in any orientation
LOI >27%			LOI <21%
Possible usage:	LOI 25 — 27%	LOI 22 — 25%	Possible usage:
PC boards	Possible usage:	Possible usage:	Normally prohibited
Relay dust covers	Connectors	Low mass, low usage items such as labels.	Need a waiver for exception, e.g. when polyacetal or poly-methyl methacrylate must be used.
Capacitor housings	Latches		
Rack covers	Gears		
Rotary switches	Pulleys		
Wire insulation			

British Post Office concept
LOI = Limiting Oxygen index

flame itself) are the norm, volatile fire retarding additives may diffuse out of the material over a period. This would leave the base material ultimately unprotected. So fire retardant material stability must be related to end use conditions, as well as to a fire hazard.

As part of the relative geometrical positioning of flammable components within a product, highly combustible materials should not be sited adjacent to high-voltage power supplies, and where bundles of wire or cable pass through bulkheads, the residual gaps should be filled with fire-stop material. Components liable to overheat should be located in areas of good ventilation and not placed below other materials which are more easily ignited. Provision of metallic fire barriers around sensitive areas may not normally be economically feasible unless, of course, it is required anyway for electrical screening. Such barriers would usefully contain the heat-damaged area in the event of a fire, although they would not limit damage remote from the corrosive smoke.

If the worst has come to the worst, and a product catches fire through internal faults, the best that could have been done would be to ensure that cover plates, housings and boxes are able to withstand at least a small amount of heat, so that they can effectively quench the fire by oxygen starvation, assuming that fresh air is not admitted until materials have cooled down. An alternative is to build-in protective BCF gas-quenching devices. This concept and other aspects of configuration control to reduce flammability in equipments is visualised in Figure 8.3. Remember that the equipment design concepts offered here do not

Fig. 8.3. Configuration control of equipment flammability. (Copyright: 1984 Standard Telephones and Cables plc.)

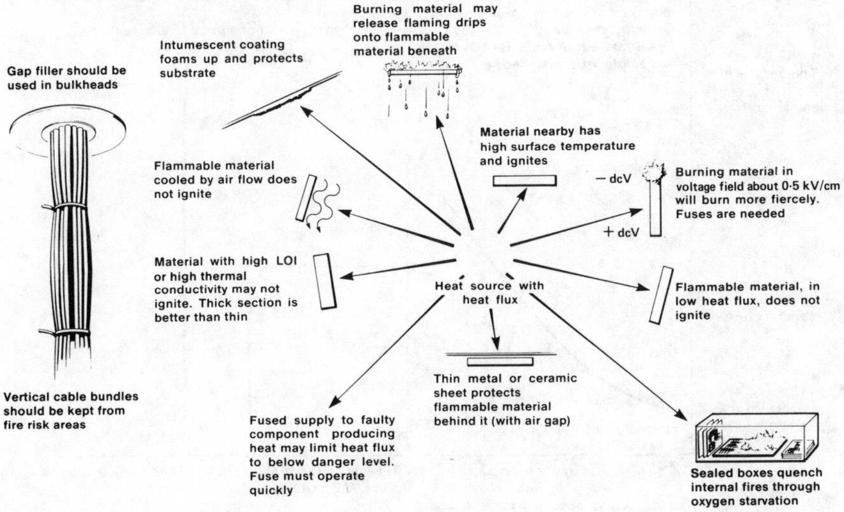

Gap filler should be used in bulkheads

Intumescent coating foams up and protects substrate

Burning material may release flaming drips onto flammable material beneath

Flammable material cooled by air flow does not ignite

Material nearby has high surface temperature and ignites

$-$ dcV

$+$ dcV

Burning material in voltage field about 0·5 kV/cm will burn more fiercely. Fuses are needed

Material with high LOI or high thermal conductivity may not ignite. Thick section is better than thin

Heat source with heat flux

Flammable material, in low heat flux, does not ignite

Vertical cable bundles should be kept from fire risk areas

Fused supply to faulty component producing heat flux to below danger level. Fuse must operate quickly

Thin metal or ceramic sheet protects flammable material behind it (with air gap)

Sealed boxes quench internal fires through oxygen starvation scale up linearly in size; they may not work in volumes as large as a medium-sized room.

8.5.3   *Materials selection*

Mention has already been made of the natural flammability of organic materials and the sense of choosing those containing additives which delay the onset of ignition in the presence of a heat source. Although, for example, polyethylene will burn quite freely in its normal virgin polymer state, additives can be incorporated to delay the onset of ignition when the surface is subjected to a flame or an equivalent temperature. This delayed onset of ignition can buy invaluable time if the source flame is likely to be quenched anyway in a reasonably short period. If, however, the source flame continues, then ultimately the fuel material will heat to ignition point and continue to burn; once burning, it will, if totally consumed, emit the total heat content of the plastic irrespective of the additive present. However the additive may play its part after ignition by endothermal decomposition, thus reducing the burning rate, so that the total burning time is increased but the immediate heat flux (heat emission rate) is reduced. This means that the fuel material is more likely to burn out without affecting adjacent materials so badly. These concepts of delayed ignition and total burning must be clearly understood for guidance when decisions have to be taken. The different situations are illustrated in Figure 8.4.

Fig. 8.4. Profiles of ignition and subsequent combustion. (Copyright: 1985 Standard Telephones and Cables plc.)

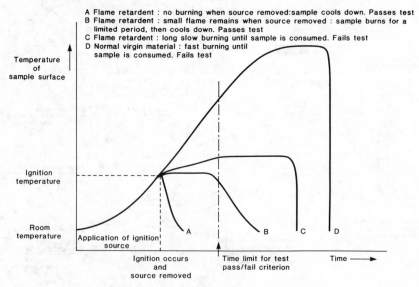

Some shortcomings of delayed ignition additives in materials should be recognised, so that proper discussion on the envisaged use can be held with the materials supplier. An additive may, under various circumstances, affect the moulding shrinkage of thermoplastics (a vital design parameter), and may also cause mould corrosion during long production runs, with consequent risk to surface finishes. The bulk mechanical, electrical and thermal properties might be changed, as well as optical transparency, specific gravity or water absorption. Additionally, surface properties of a material including these additives may be different, in the sense that a volatile additive may diffuse out, leaving the surface less protected against ignition than had been intended. Additives may be washed out or changed by solvent cleaning of the part, or destroyed at soldering temperatures and, furthermore, vapours from them may affect adjacent electrical contacts or corrode dissimilar metal couples nearby. These and other problems may not be so important once their existence is recognised, and the designer makes a judicious selection to reduce their effects.

Discussions with materials suppliers will always raise the question of limiting oxygen index, quoted as a magic number, which designers are exhorted to use when selecting materials. This index is a number arising from a specified flammability test, and represents that percentage of oxygen, in an ambient atmosphere around a recently ignited test piece, which enables it just to continue burning for a specified time. A moment's thought will show that any material which is very easily ignited at low temperatures, for example paper, must have an oxygen index at or below the value pertaining to normal atmospheric air, that is approximately 21 %. So materials with limiting oxygen

Fig. 8.5. Limiting oxygen index affected by test temperature. (Copyright: 1984 Standard Telephones and Cables plc.)

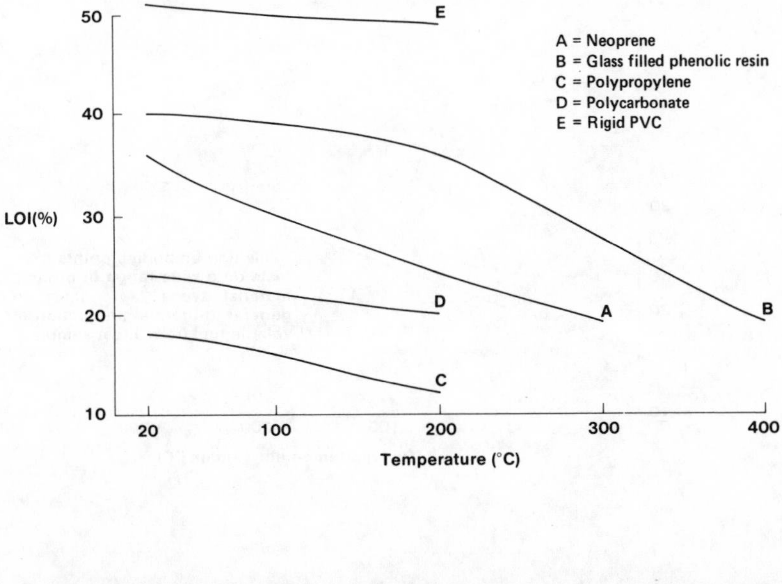
A = Neoprene
B = Glass filled phenolic resin
C = Polypropylene
D = Polycarbonate
E = Rigid PVC
LOI(%)
10
20
30
40
50
20
100
200
300
400
Temperature (°C)
A
B
C
D
E

indexes of up to 21 % will ignite very easily in air, whereas the higher the number the more the sample has to be heated, or the more the ambient oxygen content has to be increased, before ignition can begin. Obviously, if a material has a fairly high limiting oxygen index value, such as 40 %, it may happen that a burning sample of it will cease active burning once the initial heat source is removed.

At one time, limiting oxygen index numbers were offered as a panacea for all flammability ills, and customers frequently specified that materials of construction have values not less than 25 %, 27 % or whatever appealed to them. That was before it was recognised that the limiting oxygen index value, as recorded in the standard test, is itself temperature- sensitive. Figure 8.5 shows how, as the temperature of a test piece increases, the limiting oxygen index value decreases. Once the value reaches 21 %, then the heating source is able to initiate ignition. Tests on a very wide range of organic materials, of which those in Figure 8.5 are merely exemplary, provide the result shown in Figure 8.6 which is indicative only. Again we come to the obvious conclusion that the more a substance is heated the easier it ignites. A simple domestic example is illustrated in Figure 8.7, which explains why a coal fire can never be lit by a match alone, but needs the additional energy output of burning paper and sticks to heat enough coal to the required temperature, in the face of significant thermal conduction, for surface ignition to take place. In the standard limiting oxygen index test, the value obtained for coal is 44 %, effectively but

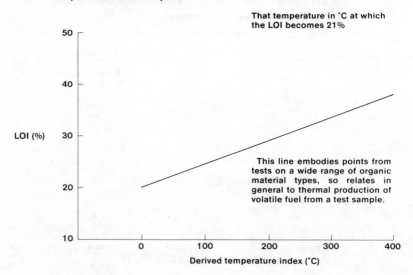

Fig. 8.6. Derived temperature index. (Copyright: 1984 Standard Telephones and Cables plc.)

That temperature in °C at which the LOI becomes 21%

This line embodies points from tests on a wide range of organic material types, so relates in general to thermal production of volatile fuel from a test sample.

LOI (%)

Derived temperature index (°C)

misleadingly indicating low flammability. This, at first surprising, result accords with everyday experience, and illustrates how the phenomenon of materials flammability needs thinking about. This value for coal should be compared with endeavours, not so long ago, to specify plastic materials with a limiting oxygen index of, say, 27 %, as a way of very safely controlling the flammability of their electrical equipment!

The limiting oxygen index method provides a series of test result numbers which can, however, be used as an aid to short-listing candidate materials already selected for other reasons. It should be considered only as an aid to other selection processes, and not a useful end in itself. In general, the temperature index for a useful range of materials is not, at present, sufficiently well documented for it to be a significant factor in the process of selection, although its existence should be recognised.

8.5.4 *Choosing a suitable test method*

Some mention has been made in previous paragraphs of selecting test methods. Their numbers at present are legion. For the purposes of this book, a few have been selected from the more broadly recognised world sources as being suitable for application to a very wide variety of tangible products in which materials are important. All flammability tests depend upon using a small range of heat sources, which are diagrammatically illustrated in Figure 8.8. It can readily be accepted that a glow-wire heat source simulates a red-hot, wire-wound resistor operating in a fault condition, perhaps resting against a

Fig. 8.7. Ignition of coal. (Copyright: 1984 Standard Telephones and Cables plc.)

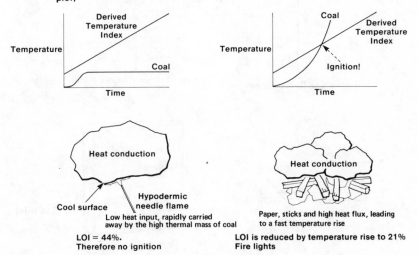

printed wiring board or a plastic spacer. The hypodermic needle flame plausibly represents any small jet of flame which, for example, may come from the end of a defective capacitor under overload conditions. Both of these sources are of low heat content, and in practice would be likely to offer a hazard for only a short time; this reality is reflected in the relevant tests where heat-source contact times are measured in seconds. The larger gas flame heat source is used for testing whole products and simulates the effect of an already existing minor (but fierce) fire within another part of the product or equipment. The radiant panel, on the other hand, is used to simulate a very large heat source for assessing the hazards of flashover fires; in these the temperature of the whole product rises as one, so that ultimately each surface catches fire at the same time, having reached its allotted position on the temperature index curve, see Figure 8.6. Some reasons for the wide diversity of test methods can be seen by examining Tables 8.9 and 8.10, showing which tests are best suited for different types of materials, for different physical forms of materials, and for components, assemblies and complete products. Many of these tests have built into them suggested acceptance criteria, by which the test material or sample may be considered to have met the test requirements. Here, obviously, is an opportunity for discussion with customers on choice of tests to take account of special

Fig. 8.8. Heat sources for flammability testing. (Copyright: 1984 Standard Telephones and Cables plc.)

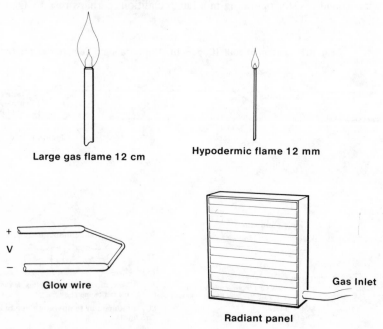

Large gas flame 12 cm   Hypodermic flame 12 mm

Glow wire   Radiant panel   Gas Inlet

Table 8.9. *Flammability test methods suggested for different classes of materials*

Test Method	Thermo-plastics	Thermosets	Plasticised resins	Reinforced plastics	Cellular plastics	Fabrics	Elastomers	Paints and lacquers
ISO 181-1981	X							
ISO 871-1980	X	X						
ISO R1210-1982	X	X			X			
ISO R1326-1970	X	X						
IEC Publication 249-1 (1968)				X				
IEC Publication 695-1	X	X						
ASTM E162-1983						X		X
ASTM D568-1977			X					
ASTM D2633-1982	X	X		X	X		X	
ASTM D2863-1977	X	X		X	X		X	
UL 94-1981	X				X			
UL 478-1981	X	X						
BS 4735-1974					X		X	

Table 8.10. *Flammability test methods suggested for different forms of sample*

Test method	Forms of sample							
	Extrusions	Fibre	Mouldings	Powder and pellets	Surface lacquers	Components	Assemblies	Complete product
ISO 181-1981	X							
ISO 871-1980	X	X		X				
ISO R1210-1982	X	X						
ISO R1326-1970	X	X						
IEC Publication 65-1976								X
ASTM E162-1983					X			
ASTM D568-1977	X	X						
ASTM D757-1977			X				X	
ASTM D2843-1977	X							
ASTM D2863-1977	X							
UL 94-1981	X							
UL 478-1981	X						X	X
BS 415 Clause 20.1 1979						X	X	X
BS 4735-1974	X							

conditions in the operation of a new type of product. Here too, the designer might well remember that although such tests are carried out under laboratory conditions which are nominally absolutely identical in all cases, in reality they may show systematic differences between laboratories. For example, in different parts of the world, or in different seasons, even the relative humidity of the test room can, in some tests, affect the result and may make all the difference between a pass or a fail for the same material. The inherent variability in replicates of these tests and of other factors (not all of which are even now fully understood) should inculcate in the designer a healthy scepticism, and a mental attitude which does not allow him to accept unquestioningly the opinions, or even the professional utterances, of the 'experts'.

The fact that in some instances, a test listed in Tables 8.9 and 8.10 may be paralleled by another one in the same list, is not necessarily a case of exact duplication. The simple situation that a national standard test is derived directly from an international standard is not, of course, reflected in these tables. However, in some instances there is a correlation yielding a similar ranking order for test samples measured by each of some pairs of tests. However, all the tests listed are unique when taken in their entirety, and should be considered as a basic palette of methods, from which the designer should consider specifying.

8.5.5   *Extrapolating test data to real life conditions*

It is assumed that the designer has by now selected his materials on the basis of their performance in relevant flammability tests, and with due recognition of the hazards and pitfalls of relying upon a simple limiting oxygen index figure as specified by a customer (or an employer). We will assume further that simulated performance tests have been made on representative individual components or sub-assemblies, using test methods which employ realistic heat sources representing any fire-causing fault conditions which can be foreseen.

Now we must face the fact that there is often a lack of correlation between laboratory-test results and real-life conditions. Despite sterling efforts by official and other laboratories in many countries it is still not absolutely certain that one can, from a small test mock-up, predict real-life fire conditions. About the best that can be done here reflects room fires, where some correlation is observable. However, the experts are quick to point out that minor changes in real life conditions can influence fire spread and other parameters, so that it is almost never possible to predict absolutely reliably how a fire, once started, will spread in a large area.

One of the difficulties is that test samples do not represent the correct area to volume ratio, as do the full-sized artifacts. A difficulty lies in simulating the scalar quantities of heat content, specific heat and thermal conductivity in a test specimen, although computer modelling of room fires is beginning to be done.

Furthermore, laboratory tests are often carried out for standardisation purposes, using a limited heat source applied to a cold sample edge, whereas in most real fires the heat arrives as a flux of radiation impinging upon an already hot surface. A bunsen burner flame, both in size and in temperature profile, does not adequately reproduce the lower temperature flames which might occur in a real-life fire spread depending perhaps upon burning wood or card as initial fuel.

8.5.6   *Smoke emission*

While not of prime importance within the context of this book, the reader should be reminded that, when any material burns, some air-borne combustion products are always emitted. In the case of a very clean coke fire, the emitted product may be the invisible, tasteless carbon monoxide gas, whereas with less complete combustion, aerosol particulates containing tars and aromatic compounds centred on soot nuclei will fill the air with unpleasantness and obscuration if not toxicity. In more extreme cases, often involving man-made materials, other chemicals can be produced during combustion which are irritant to the eyes or respiratory system, or which may show insidious effects, as in the case of cyanides. In all these cases the rate of smoke generation is more important toxicologically than the total quantity produced.

Sometimes smoke acts detrimentally towards parts of a product not directly affected by the heat of a local fire, or it may travel around the room and impinge upon other products or equipments innocently engaged in their business, yet in imminent danger of suffering corrosion from combustion products. (It is for this reason that fires in telephone exchanges are so much feared by insurance companies.) Although there is considerable interest in, and development emphasis on, the problems of toxic smoke in public transport vehicles such as aircraft, Anon. (1983*e*), less emotive, but for present purposes technically more important, publications on controlling smoke and toxic gas evolution are regularly produced. That by Grayson, Hume & Smith (1982) is a good example. Here as for some other subjects, the reader should not expect to find, read and understand this literature himself. He should be aware that it exists, and that plastic materials suppliers should be reacting to the results and the recommendations presented therein.

8.5.7   *The problem of qualification testing of products*

The reader will, by now, have digested the message that safe and reasonably certain knowledge of the fire behaviour of his product will only be obtained by testing the product itself (or by a sensible mock-up) using a fire source approximating that which would be envisaged as the main cause of ignition under probable fault conditions.

The failure of laboratory flammability tests to correlate perfectly with real-life conditions should justifiably prevent the informed layman from judging relative fire hazards from a single test result. An example of this, some years ago, was a case in which six different European Countries, through their test laboratories, carried out round robin trials on the combustion ratings of 24 materials, and themselves interpreted the results. These showed poor correlation of the relative ranking orders found by the various laboratories, the most extreme example being phenolic foam which was rated the most combustible in Denmark, but least so in Germany!

8.5.8   *and finally …*

Perhaps this chapter can best be summarised for the designer, who may well, at this point be totally confused by the complexities of flammability, by referring him to Figure 8.9. This shows the three ingredients needed to start a fire: available fuel, the correct proportion of oxygen, and sufficient heat content at an adequate temperature to cause ignition. It is obvious that there are extreme limits to these three parameters, outside which there may be plenty of heat, but no fire. The reader could, as an exercise in understanding, try to indicate on this diagram those factors, such as limiting oxygen index values, which play their part in fire initiation and growth. However, the diagram shows the challenge that the designer must overcome. He has to arrange his design such that during service life and under all foreseeable operating conditions of his product, the central zone of fire condition is never intruded upon.

Fig. 8.9. Requirements for a fire. (Copyright: 1984 Standard Telephones and Cables plc.)

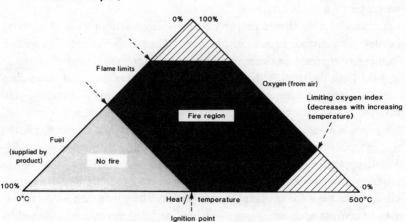

9

Finishes and coatings as protective systems

What is it about an engineered product which appeals to a potential customer at first sight? It is arguably true that shape and surface appearance are persuasive factors, even for products outside the consumer goods industry. While any customer, be he private or industrial, will purchase an item primarily for his expectations of its technical performance, nevertheless, much of the joy of ownership lies in the outer surface features. An obvious example here is a motor car for which, to the uninitiated, the difference between a desirable car and a run-of-the-mill vehicle is often only signalled by differences in shape, surface gloss, choice of colour and general decor. The designer of engineered products is, of course, aware of this, and should not be beguiled by subjective considerations, but nevertheless may find he lacks a general overview of finishes and coatings as protective systems with which to enhance his brainchildren. This chapter does not claim to offer a comprehensive treatment, but will discuss the more important aspects of surface finishing of which the designer should be aware. So the purpose of a finish is to be decorative, functional or to permit repair of a worn surface.

Surface finishing should properly be considered as a process of applying a system to a surface in need of treatment. The common belief that surface finishing is confined to cleaning a surface and applying a coat of colour is very far from the truth in the modern world. Whether the finish is an electroplated layer, an organic paint film, a sprayed ceramic or what-have-you, there are general principles which must always be considered before an effective result is achieved. The substrate material, which may be metal, plastic, or even timber-based, has first to be prepared to ensure adhesion between subsequently applied surface layers. In the case of a metal, there may then follow a conversion coat comprising a closely adherent compound of the metal, which will form the basis for an organic-based primer. There may then follow undercoats to provide adequate colour, body and surface smoothness, followed by a top coat intended to provide the correct level of gloss, texture, colour, and possibly any required abrasion resistance or other performance characteristics. Obviously, these stages are not all needed for all types of substrate. Thus, for example, a metal may be cleaned to accept a single layer or multilayer of an electrodeposited material, which then comprises the finishing system.

Having recognised that the finish on an article has to be considered as a system, thought must then be given to the requirements it must meet. In general, these are four in number: finishes may be applied for protection; to enhance the servicability of the basis material; to add commercial value to the article; and for aesthetic reasons; and while each of these four functions could stand alone, some contribution from the other three will always attend the main requirement to be met.

Finishing systems are most often applied to substrates to protect them from the environment. A paint film may confer resistance to air, water, organic liquids or chemicals, as well as improving mechanical properties through increasing the hardness or abrasion resistance. Other paint-type coatings may be fire-retardant to protect combustible substrates, or may be applied to porous surfaces such as concrete or plaster to facilitate later cleaning. Apart from organic-based finishes, metals are protected by the application of metallic coatings, by chemical conversion coatings, by applied ceramics or by the use of polymer coatings. These alternatives are discussed later in this chapter. Plastics may also be painted for protection against strong light, water, salts, or other adverse environmental conditions.

Of the metals used for engineering purposes, primarily iron-based materials need painting or surface protection because of their inherent tendency to corrode in damp air. However, the act of joining two different metals may lead to corrosion of one of them, as implied in Table 9.1, which illustrates one version of the well-known galvanic series of metals. In this series the upper members are those most likely to be corroded, whereas those at the bottom of the list are more stable towards corrosive influences. In this table, the terms 'active' or 'passive' respectively indicate whether the metals have a clean, reactive surface, or whether they have been passivated by natural oxidation, chemical or other special treatments. That such treatments may be crucial, in controlling galvanic corrosion between metal pairs, can be seen in the cases of nickel or chromium − nickel − iron, which have very different positions in the galvanic series depending upon the presence or absence of this treatment.

Surface finishing systems are sometimes applied to impart a specific technical function to the substrate. For example, paints may be used to confer an electrically insulating surface upon metals, or in metal-filled form produces an electrically conducting surface on plastics, enabling statically generated electrical charges to leak away safely. Plastics are painted to improve abrasion resistance, and paint films have on occasion been used to act as a diffusion barrier against unwanted contaminants.

Commercial value, of course, can be added by properly painting industrial equipment, road vehicles or civil engineering items such as bridges, to which surface protection is applied to reduce later maintenance costs. In general, such finishes are not expected to meet the aesthetic appeal requirement usual for consumer goods.

In the consumer arena, surface finishes, and applied colours and textures in particular, are used to maximise customer interest at the point of sale. Correct choice of colour, and the balance between finished and metallic surfaces, is

Table 9.1. *The galvanic series of metals*

Corroded end (anodic, or least noble)

Magnesium and its alloys
Zinc
Pure aluminium
Cadmium
Aluminium – copper type alloys
Iron or steel
Cast iron
18-8 chromium – nickel – iron (active)
18-8-3 chromium – nickel – molybdenum – iron
Lead – tin solders
Lead
Tin
Nickel (active)
Inconel (active)
Hastelloy C (active)
Brasses
Copper
Bronzes
Copper – nickel alloys
Nickel (passive)
Inconel (passive)
Titanium
18-8 chromium – nickel – iron (passive)
18-8-3 chromium – nickel – molybdenum – iron
Hastelloy C (passive)
Silver
Graphite
Gold
Platinum

Protected end (cathodic, or most noble)

very important from an aesthetic point of view, as is an appropriate choice of gloss level on the outer surface. Colour may be applied through painting a surface, or it may show through from an underlying coloured surface subsequently protected by a clear lacquer coating. Fillers in the paints can be chosen to show special optical effects such as metallic lustre, pearlised or hammer finish or fluorescence. Increasingly important for aesthetic purposes is the painting of those plastic parts, used in a composite construction, which must look identical to the rest of the product and which may perhaps be manufactured from metal, for example car bumpers.

It is frequently asserted that the final manufacturing process for a product, which is usually application of a surface finish, was the last factor to be considered in a design but nevertheless is the first item on the product to suffer when cost-reduction exercises are undertaken. It should be obvious that when developing a design the surface treatment to be applied should be decided at a very early stage so that appropriate adjustments to the engineering of the product may be made. The designer must be fully aware of the production processes by which the object is to be made, and these include not only his in-house current methods for applying the finish, but the handling processes employed both before and after the finishing system is applied. A simple example: it would not be sensible to select a stoving paint system for finishing a very large object if large curing ovens were not available to the company. Table 9.2 suggests a checklist of factors to be borne in mind when considering the finishing aspects of a design; use of this will avoid some of the problems which have caused difficulties in the past.

As an example of product shortcomings likely to occur through ignoring the technology of the finishing process, we can imagine coating a metallic piece part with paint. If a liquid paint is applied by spraying, then it must be accepted that the paint film will be thinnest at sharp edges and points, and will build up in threaded holes. If electrostatic spray painting is chosen in an attempt to eliminate some of this thinning, then, perversely, paint will build up at sharp edges and discontinuities as it is here that the electric charge density is greatest so that the paint tends to be attracted the most. The amount of build-up under these conditions will depend upon the evaporation rate of the solvent in the paint and the rate of viscosity increase of the drying paint film during subsequent oven stoving. If, on the other hand, solvent-based paint is applied by a dipping process or by an electrophoretic process employing water-based paints, there is then the potential for applying paint uniformly all over the surface. However, in practice, use of ordinary liquid paints will lead to thickening at points where there is drainage, and re-entrant curves and blind holes will probably not be coated unless the design provides vents in the part to permit free escape of air when it is immersed in the paint tank. Such difficulties are largely overcome by the electrophoretic process if air vents are provided. If plastics are painted then one must consider whether threaded inserts are to be used, as these should, if possible, be fitted after the painting process. It goes without saying that in all cases the places at which the part will be supported during painting should be carefully selected so they will not show on significant surfaces.

Table 9.2. *Check list for a designer considering the product finishing phase of a design*

Define the planned product life

Identify the static or dynamic environment in service

Identify those finishing systems which will provide protection during product life

Assess the relative costs of the systems

What effects may the chosen system have upon the product design?

Can the following requirements be met?

    Appearance
        Colour, texture, gloss

    Corrosion protection
        Against the foreseen environments

        Is there any possibility of bimetallic or crevice corrosion?

    Technical aspects
        Must the finish be present or undamaged despite welding, soldering or adhesive operations?

        Can it withstand stress, bending, impact?

        Must it withstand a heat treatment?

        Can it conform with dimensional requirements in the finished part?

        Must the finish meet any electrical or magnetic requirements?

        Can the finish be applied without marks from process support points being visible?

9.1　Characteristics of surface coatings

Finishing systems must fulfil a number of widely different expectations in service, and must be chosen carefully, bearing in mind the consequences of finishing system failure. A familiar domestic example is the application of oil-based paints to wood which, with variations of temperature and humidity, changes dimensionally in a way that is reflected by the flexibility of the paint film. However, other factors eventually cause this film to become harder and more brittle, with cracking and the need for a further painting job as end results. Extreme flexibility of a surface finish can be achieved by application of plastisol dip-coats, but these are mainly suitable for small articles which can subsequently be oven-dried to harden up the applied layer.

When metals are to be protected against corrosion, a variety of choices is available; for example a rigid barrier coat of non-corrodible metal may be applied to a the substrate. A case in point is nickel-plating over a copper strike (thin film) applied to an iron or steel article. All is well so long as the barrier coat is not breached, but corrosion can occur if it is, when the underlying iron becomes directly attacked by moisture passing through the surface defects. On the other hand, an electroless nickel deposit on a plastic article will, if it cracks, not permit corrosion, but may have other aesthetic disadvantages. Should the possibility of using a plated barrier protection coat be unthinkable, then one may have recourse to a process where a protective metal is sprayed onto a corrodible metal substrate. This is generally done when aesthetics are not important and reliable service is a prime requirement, as in the case of the surface protection of under-sea repeater housings for submarine cable systems.

Economics may demand that, in the case of a product with an expected short life, a sacrificial metal coat should be applied to a ferrous substrate, the coating protecting only for the required lifetime period. Specifically, one can recall how electrodeposited zinc on iron forms a good protective system, preventing the iron from corroding, because the zinc is sacrificed during the corrosion process until it begins to fall away as a white powder. At this stage, iron corrosion occurs on the bald patches at the normal pace.

Sometimes, and especially in high technology applications, the surface of a product itself performs the prime technical function, while the interior is expected only to provide form-stability and support to the surface. A finishing system of this type would comprise ceramic layers applied by flame or plasma spraying on jet engine parts, fan blades and bearings.

Conventional organic-based paints, basically organic film-forming resins pigmented to the desired colour, usually provide a barrier type of protection. Under normal conditions, they form an adherent, resistant, conformal barrier between the environment and the structural material to be protected. Successful protection, however, demands that the coating be applied, and maintained, as a continuous film. On the other hand, the sacrificial protective action of zinc-rich paints or of zinc plating does not necessarily depend upon absence of pin-holes, as protection is still provided at these points by electrolytic action.

Zinc applied by the hot galvanising process forms intermetallic alloys with an iron-based substrate; these alloys have intrinsically corrosion-resistant properties, so that a galvanised article is much less ready to sacrifice its surface under impact and abrasion conditions. The finish is technically more effective than an electroplated zinc layer, aesthetically not so pleasing, and of greater cost.

The cost of surface finishing systems is, naturally, of prime importance in any industrial activity. In an age not noted for the price stability of commodities, it would be pointless to quote specifics for costs of materials, or even to quote relative ranking orders for costs. It must suffice to warn that the product designer should be aware that the cost of his finishing system is in part based upon the costs of preparing the substrate to receive the system, the cost of the materials used in the system, and the cost of the application process itself including post-processing activities such as surface polishing. Sadly it tends to be true, however, that in many enterprises finishing costs are amongst the least understood and worst documented of the total manufacturing operation, so the designer seeking information from his colleagues must be prepared for at least minor disappointments, such as receiving misleading, but well meant, data. This result can be avoided if the problems are discussed with experts on finishing.

9.2.   Selection of finishes

Table 9.2 offers a checklist for consideration when selecting a finishing system for a product, and a little has already been said in the previous section on design for finishing. At the present stage, the designer has to be rather more specific in his thinking to focus upon a short list of suitable systems.

Factors in selecting a finishing system can range from a simple recognition of the nature of the substrate to be finished and of the service conditions to be met, through to an adherence to predetermined company image (which may prove to be either a constraining or helpful effect). Usually the company image will have been developed before the design of any particular product, and so should not be difficult to achieve in the light of previous experience. However, the very existence of a company image or style of product finishing may result in a limited choice of finish application processes being economically available. Part of this consideration must be the probability that the finishing system chosen will achieve a specified technical performance, and that it will yield aesthetic satisfaction. If most of the company's products are of similar nature, then aesthetics will no doubt take care of themselves. If, however, a sub-component in a product is to be designed afresh to achieve better technical capabilities, then it may be that special surface finishes, not within the manufacturing capability of the host company, have to be selected. In such a case, the designer should not hesitate to seek specialist advice. An example of this situation would be the deposition by specialised processes of ceramic or intermetallic coatings onto engineering surfaces to improve wear resistance while the rest of the product stays essentially unchanged.

Before proceeding to a review of surface finishing processes and of the currently most important finishing systems, we should consider, very briefly, the nature of the four main finishing system families.

The most common, oldest and simplest to apply is the organic-based finish which has existed since Roman times in the form of vegetable drying oil vehicles with pigments, and latterly anti-oxidants, accelerators and a number of other proprietary aids to improve ease of application and performance.

In another family, plastic layers are deposited as protective films, ranging from the thick plastisol type covering familiar in bicycle handlebar grips, to closely adherent, thin, non-stick, polytetrafluoroethylene surfaces familiar in good quality domestic frying pans.

Surfaces protected by metals can be seen in all walks of life. Metals can be deposited in thin or thick films by evaporation, by liquid spraying from heated powder or wire, or by deposition from solution onto conducting substrates. (Plastic surfaces can be given a conducting function for this purpose.) In addition, metals can be introduced into a substrate metal surface by using solid diffusion (pack) techniques, or by diffusion from a melt, as in the hot-dip galvanising of iron. Metal surfaces are also modified by heat-treating the substrate in an appropriate environment, for example by the nitriding and chromising of steels, or by bombarding the surface with atomic particles as in the case of ion plating.

Protective layers are formed on a substrate by chemical conversion techniques which, in the case of some metals, produce an inert inorganic compound as an adherent and protective layer. Ceramics can be deposited onto many surfaces by wet electrophoretic techniques or by spraying at high temperatures.

Thus it is obvious that there is a generous palette of methods available to the designer who, if not sufficiently expert himself, should be prepared to discuss his problems with a knowledgeable consultant.

The possibilities of finishing systems are legion. Even if the substrates are restricted to the generic classes of metals, plastics, natural products and ceramics, there still exists a very wide range of manipulative processes for applying the various candidate finishes to them. In the context of what is meant to be a fairly brief text, it would be over-stressing the subject to provide compendious tables of the names of processes, together with advantages and disadvantages of each; yet an attempt to restrict discussion to the more commonly known finishing systems might not only lose the interest of the

Table 9.3. *Some less familiar surface finishing processes*

Gas flame spraying	Metal wire or ceramic powder is fed into an oxyacetylene flame and propelled to the workpiece by compressed air
Detonation spraying	Metal or ceramic powder and an explosive gas mixture in a chamber are ignited by a sparking plug. The explosion product and the melted powder are ejected onto the workpiece
Arc flame spraying	An arc is struck between two metal wires, and the metal vapour is directed to the workpiece by compressed gas
Glow discharge	A gas discharge at low pressure polymerises organic gases present, to deposit a pin-hole free polymer layer onto the workpiece
Plasma flame spraying	Powdered metal is propelled through an arc-driven gas plasma, to impinge on the workpiece
Case hardening	Usually applied only to steels, a generic heat treatment process to increase surface hardness. Carburising, nitriding, boriding, chromising, aluminising and 'sulphurising' are heat treatment methods in which the steel surface receives the appropriate element to increase surface hardness or to reduce wear
Hot-foil stamping	Preformed coloured or textured foils are impressed into a plastic surface
Sublimation dyeing	A plastic article is held in the vapour of a dye, and absorbs the colour into its surface
Dip dyeing	Immersion of receptive plastics, for example nylons, into a dye solution, which is surface-absorbed

knowledgeable reader but might bypass entirely the interests of those with special needs. In order to avoid presenting a glossary, and also to avoid pre-empting the discussion in Section 9.4 of this chapter of some individual methods, a brief description is provided in Table 9.3 of a few of the less obvious finishing methods, of which the designer should be aware. Reading this table will not give him technical knowledge of unfamiliar processes, but he will at least have a conceptual aid to understanding that he can enlarge upon by reference to literature or from expert sources.

In addition, and to provide some framework for a very complicated picture, Table 9.4 lists, by name only, the main processes by which liquid paints are applied to a variety of substrates.

Emphasis must again be put on the paramount need to consider substrate preparation before any coating process is applied. In the case of metals, this can involve removal of the grease by organic solvents or by detergents, chemical

Table 9.4. *Summary of liquid paint deposition processes*

Spraying
    Compressed air
    Steam (to atomise the paint)
    Hot spray (solvent evaporates to atomise paint stream)
    Electrostatic air-less (for better surface coverage)
    Electrostatic (to a conductive substrate. Can now use water-paints)

Dipping

Electrophoretic (paint 'electroplated' from an aqueous suspension)

Tumbling (small metered amount of paint used in a container)

Curtain coating

Flow coating

Knife coating

Hand brushing

Machine roller coating (often for steel coils)

Hand roller coating

cleaning and descaling to remove pre-existing corrosion products or other process chemicals from the metal surface, followed by derusting and possibly some form of surface shot blasting or similar blast cleaning to provide a mechanical key which is, at the same time, still chemically clean. The ideal finishing process is fully in-line from start to finish of the sequences but in reality a workpiece is likely to be static for some while after a sequence before a subsequent process is applied. The delay may necessitate a further pickling operation, often followed by an anodising or a phosphating treatment before the metal is given its first coat of priming paint.

Some further explanation of the consequences of paint deposition processes listed in Table 9.4 is probably justified. Each method involves its own plant investment, for which conventional air spraying of paint is usually the lowest. Air spraying is subject to a number of refinements insofar as the liquid paint may be hot or cold, and the gun can be manually or automatically controlled, or even possibly driven by a computer-controlled robot to ensure exactly repeatable coverage of a series of large or complex objects. Air-less spraying can also be operated manually or automatically, is less wasteful of paint but more expensive in the plant needed for it. Electrostatic spraying can be used equally with solvent-based paints or water-based paints, and may rely purely on electrostatic formation of the paint mist, or on compressed air. Electrophoretic coating is, in concept, somewhat similar to electroplating, being suitable for depositing very uniform films over the whole wetted surface of an immersed object. However, the design must take care of drainage holes to allow air out and paint in, otherwise box sections will not become internally coated.

In fluidised-bed powder coating, the article to be painted is first heated to fuse the paint powder subsequently deposited upon it from the fluidised bed. When operating the electrostatic powder coating process, the powder is applied dry to the article, clinging to it by electrical attraction, and then needs a subsequent oven stoving to melt it into a coherent skin. Many of the application methods rely on the paint being air-dried, but for professional purposes, to achieve speed and good service performance, it is more often high-temperature dried and cured; latterly paint curing by application of ultra-violet or infra-red radiation, or even by high energy electrons has been successfully effected.

At times, a surface texture or other requirement will decide the application method. Thus a spatter-finish (in which the paint forms small mounds on a smooth substrate to provide an attractively rough surface) cannot be obtained by dipping or by electrostatic spraying processes, but only by using compressed air or air-less gun spraying methods. Very small piece parts may be uneconomic to paint individually, and may need processing by a batch, barrel or centrifugal process. On the other hand, very large parts may be chosen for stove-finish painting but are too large to fit available stoving ovens. In such cases possibly an infra-red or ultra-violet light curing system should be adopted.

Painting on plastic substrates offers another challenge. When stoving can be carried out, this will usually be at rather low temperatures; consequently drying times may be quite extended, a clean area being needed to avoid dust pick-up onto the still-wet paint film. Most plastic injection mouldings are designed not to have very large flat areas for aesthetic and technical reasons, and also because they are not very receptive to air-less spraying to provide a high quality finish. The majority of plastic painting thus still uses air-driven spray guns, although roller coating and flow coating have been found suitable for plastic surfaces which are smooth and flowing without large changes of curvature. Because of the large proportion of solvent usually present in paints, dipping and flow-coating with thermoplastics is not usually advisable because of solubility and related stress-cracking risks. These techniques are, however, useful for thermoset plastics, although, of course, in most cases painting is avoided by using self-coloured materials. Manual application by brush or roller is inexpensive but seldom provides adequate quality for industrial purposes. Should electrostatic paint spray painting be envisaged for coating plastic surfaces, then a special conductive layer must be provided by chemical or other means and the part has to be jigged with an earth connection. Unfortunately, the chemical substances used are often water-soluble and residues from them limit the environmental conditions to which the article may subsequently be exposed.

9.3. Finishes for modifying surfaces

So far, this review of protective systems has emphasised the protective nature of surface finishes and coatings on a substrate. The designer has been guided through the factors to be considered when designing a surface finish, the broad characteristics of surface coatings, ways of selecting them and a brief listing of application processes. It is appropriate now to consider the surface modification of a product which is necessary to meet technical requirements or for a decorative effect.

9.3.1 To meet technical requirements

Although most surface finishing systems are applied to protect a substrate from attack by deleterious environments, there are cases where a substrate, otherwise useful for many aspects of the proposed application, nevertheless is deficient in the required properties at the surface. Metals otherwise suitable in terms of strength and toughness for, for example, sliding surfaces, may if unmodified suffer excessive wear in use, so that surface treatments are desirable. Mention has already been made in Table 9.3 of some heat treatment processes, applicable to steels, to introduce carbon, nitrogen, or other elements into the surface layer, with a consequently enhanced increase in wear resistance and, often, in toughness as well. It can be imagined that considerable expense might be incurred if a large and complex component had to be made from a specially hard type of steel just to provide extra wear resistance on some particular working surface. A cheaper solution would be to use low cost steel of adequate strength, and to apply a localised surface treatment. Localised electroplating, for example of hard chromium, is one possibility for upgrading bearing surfaces and can be applied to a variety of cast, mild or hard steels. Design information relevant to the hard chromium electroplating process, and the products' hardness and frictional parameter data, are discussed by John (1984). These coatings can be applied to non-ferrous metals such as copper and bronze, aluminium, and even to the reactive metal titanium. In addition to chromium, electroless deposition of nickel containing up to 10% of phosphorus can be employed to form hard surfaces on all steels, cast iron and aluminium, and on most conductive plastics and non-ferrous metals. Subsequent heat treatment of these coatings on appropriate substrates can significantly increase the hardness. The corrosion resistance of this form of nickel is such that it is complementary to hard chrome, resisting as it does many reagents which attack the latter metal.

In addition to these two particularly useful metals, others producing surface coatings include iron, copper, lead, silver, gold and tin. Their engineering uses range from cost reduction through corrosion protection to provision of good electrical contacts (by localised deposition of gold) and additional lubricity in bearings. Mention may be made, in passing, of the build-up and restoration of worn or mis-machined parts by use of heavy layers of nickel or chromium plating and subsequent machining to finished size.

Although carburisation of steel is usually applied to the whole surface, localised deposition of copper plate can prevent carburisation at that point, a situation exploited in some engineering designs. Tin−copper alloys are similarly used as a stop-off for the nitriding of steels.

This is not the proper place to launch into a detailed discussion of the boriding, carburising and the other heat treatments for steel surfaces, which are listed in Table 9.3. However, mention can be made in slightly more detail of the nitriding process alone, to illustrate techniques currently available. A salt-bath nitriding process for hardening steels, Satoh (1981), uses a special catalyst, while at the same time direct current is passed through the salt bath. The catalyst is pure titanium metal, which volatilises easily above 510 °C without ever reaching the liquid state. This titanium generates many fine particles of oxide which disperse in the molten salt and catalyse the process to increase the nitriding speed.

A more recent development, by Lucas Industries PLC in England, is of 'Nitrotec'. This proprietary group of processes is intended for the gaseous nitro-carburising of plain carbon and medium carbon low-alloy steels not only to increase corrosion resistance but also to produce an aesthetically pleasing finish. It is claimed furthermore that this surface treatment improves the grain struc-ture of the underlying metal so that section size can be reduced, while at the same time surface hardness is increased.

Nitriding can also be applied to aluminium alloys for sliding and rolling applications. Hioki (1980) describes a tungsten inert gas arc-melting technique to produce lamellar aluminium nitride, with enhanced hardness, over the substrate material.

Deposition of ceramic surfaces on substrates is a well-established technology, and deposition of tungsten carbide, chromium carbide, alumina or zirconia, alone or with additional lubricant metal inclusions, is reasonably well-known. Some of the methods of applying these are listed in Table 9.3; a recent exposition in detail of the detonation gun and plasma spray technologies for these ceramic materials is offered by Gill (1984). This topic is dealt with in rather more detail in Section 9.4.2.

9.3.2 To meet decorative requirements

The appearance of electroplated coatings for decorative purposes a multitude of consumer goods is familiar to everyone. A very common finish is that in which a 'flash' or thin layer of chromium is deposited onto about 25−50 micrometres of nickel plate used to protect underlying steel, as in the case of some automotive hardware. Also familiar, in the form of cheap jewellery, is a decorative gold deposit which may be as thin as 0.025 micrometres. Between these two, other materials will be met, such as tin plate for food containers, and silver for jewellery and decorative ware. Much of the decorative effect of electroplating is achieved through careful preparation of the substrate, particularly in terms of surface finish, as the thickness of the electroplated layer is generally insufficient to allow major defects to be removed subsequently by a burnishing operation. Plastics too, particularly ABS and polypropylene, are widely electroplated. This technology will be discussed in a little more detail in Section 9.4.3.

Paint finishing of all forms of material is of course very familiar, and its decorative potential is obvious: a later section of this chapter deals with organic finishes in detail.

Metals can now be coloured by a variety of methods, of which electrolytic anodising of aluminium and its alloys, followed by a dyeing process, has been familiar for many years. This can be selectively applied by a sequence of mask-ing operations.

Plastics can, of course, be painted but there are many other techniques for producing a decorative surface finish, including embossing, printing by silk screen or gravure, vacuum metallising, electroplating and decorative laminating. Each of these topics is a significant subject in its own right, and the reader is advised to consult specialist texts to learn more about them.

Chemical colouring of metals is becoming increasingly important as a decorative process. For example, conversion coatings of aluminium surfaces, using hot alkaline solutions, can generate colours ranging from grey to black and from yellow to brown, as recently discussed by John, Perumal & Shenoi (1984).

Particularly important for architectural purposes, is a new process from the Ionic Plating Co. Ltd, with the trade name 'Staycolor'. This process modifies and slightly thickens the natural passive oxide film on stainless steel, until the film thickness is similar to that of a wavelength of visible light, and the colour can be made to match adjacent electrolytically colour-anodised finishes on aluminium. Another process with the trade name 'Permacolor' from the same company involves printing coloured designs and patterns onto stainless steel surfaces using photographic techniques, with photo-resists which can withstand the colouring and hardening solutions subsequently used. Again, interference colours are obtained, but with predetermined patterns as well as colours.

9.4   Metal based surface layers

Much finishing activity today is concerned with deposition of metals or metal-based compounds to protect the underlying substrate. The following paragraphs briefly scan some of the more important techniques, not to provide the reader with in-depth knowledge, but to acquaint him with the complexity of the subject and to provide sufficient information to enable him to address his data sources sensibly, whether these be literature or people.

9.4.1   Selecting the basis metal

Surface coatings, applied to a metal substrate in order to resist corrosion and perhaps to impart better wear resistance, permit selection of a less expensive, but strong, substrate than would be the case if the coating was not used. An inexpensive substrate may compensate for the cost of a very expensive surface coating, should this be needed technically. In any event, selection of a metal surface coating must depend upon a cost evaluation of the raw materials involved and of the manufacturing processes needed to apply it, together with the likely cost of subsequent maintenance.

Coatings applied by electrodeposition, flame-spraying, hot dipping, cladding or other techniques and used for corrosion protection, include zinc, cadmium, nickel, chromium, silver and other precious metals as electroplates; aluminium, tin, lead, Monel, stainless steel and various hard facings can be applied by other methods.

Some specialist companies offer guides to their coatings, together with checklists to assist selection. These guides are often organised to first identify the function which must be met by the coating, and then to select specific materials which should meet the requirements economically. Special coatings are often offered for particular types of substrate metal; coatings produced by a gas or flame spraying process can often be composite, such as nickel−aluminium, nickel−chromium−aluminium, copper−aluminium, molybdenum−nickel−aluminium, and many others.

9.4.2 *Deposition by spraying*

Low melting-point metals, including aluminium or zinc, are deposited as coatings by a spraying process little different from conventional paint-spraying, but using a heater attachment on the gun. These sprayed coatings are often enhanced by additional coatings of primer and special sealants to achieve optimum performance. A more elaborate deposition technique involves combustion flame-spraying, in which fuel gas and oxygen burn to melt coating material fed through a gun, the material starting as wire or powder. Almost any metal, melting at temperatures as diverse as those of molybdenum and zinc, can be sprayed by the wire process. In the powder process, material is added to the flowing combustion gases which melt the particles of powder and accelerated them to a velocity of about 100 metres per second. As the melt impinges on the substrate surface, the droplets flatten and form a coherent, if porous, coating.

Requiring rather more elaborate support equipment is the plasma flame-spray gun, in which an electric arc contained by a water-cooled jacket excites an inert gas, and the plasma so produced melts powder fed into it and propelled to the substrate. Plasma spraying and the related detonation-gun technique are discussed by Gill (1984).

9.4.3 *Electrodeposition and electroforming*

Electroplating as a concept should be familiar to all readers. It comprises the electrolytic reduction of dissolved metal ions and their deposition onto the workpiece which comprises the cathode in an electrolytic cell. The metal being deposited may dissolve continuously from an anode of the same metal, or an inert, insoluble, conductive surrogate such as platinised titanium or carbon may be used as an anode. Electroplated films can be deposited onto a wide range of substrate base metals, including all manner of steels and irons, copper, brass, zinc, aluminium, magnesium, titanium, precious metals and many others, including many, but not all, alloys. It should be remembered that a significant cost factor of the electroplating process is preparation of the workpiece ready for accepting the electroplate deposit, which itself may be a complex one comprising two or more layers of different metals.

The reactivity of some metals, such as titanium and aluminium, has delayed development of electroplating processes using them as substrates until fairly recently. Steel, the most common structural metal, must be protected against rusting in almost all its applications, so the electroplating process is very commonly applied to it. Tin protects steel in the very important tin-plate and tin-can industries; a recent review of tin and its alloy coatings is offered by Chapman (1984). Zinc is very often electroplated on to steel, although for many structural engineering applications hot-dip galvanising (qv) is far more significant in terms of tonnage of product. Other metals commonly electrodeposited include cadmium on ferrous and copper alloys, chromium on ferrous, copper, zinc and aluminium alloys, nickel on ferrous, zinc, copper and aluminium alloys, and lead, copper and some alloy platings such as lead—tin. More complex electroplated alloys, for example those containing nickel, iron, chromium and phosphorus, have recently been announced by Diegle (1981). These alloys form as non-crystalline (glassy) films with interesting new properties, particularly in terms of corrosion resistance.

The availability of such a wide range of electrodepositable metals does not automatically confer unlimited design freedom to produce products of any shape which can still be finished by this process. Each electrodeposited metal has to be dissolved in its appropriate conductive solution (electrolyte), but the electrical conduction and ionic transportation efficiency of these solutions varies from metal to metal. The 'throwing power', or ability of a transported metal ion to travel round corners in the electric field, varies considerably for different metals, and the design of the cathodic workpiece has to favour uniform current distribution on it unless a wide range of deposit thicknesses can be accepted on its various surfaces. Better deposits in tubes and the like are obtained when supplementary anodes are inserted. Very recently, a ternary chrome-iron-nickel electroplating process has been offered by Permalite Chemicals Ltd under the trade name 'Oztelloy'. This deposits a thin layer of a type of stainless steel over an underlying nickel substrate and has reasonable throwing power, thus widening further the designer's options.

Until recently, it was not possible to deposit aluminium onto substrates without recourse to very complex organic electrolytes which were highly reactive and thus dangerous to operate. Unfortunately, even now the process has to be carried out largely automatically, with airlocks to contain the electrolyte chemicals. Work by van de Berg, Daerman, Krijl & van de Leest (1980) in the Philips Research Laboratory in Holland, led to selection of less hazardous solvents for this purpose. A somewhat different process has been offered by Hegin Galvano Aluminium BV (1982), opening the way to wider acceptance of electrodeposited aluminium in the future.

An obvious corollary to electrodeposition is electroforming. It can readily be imagined that if a metal is electrodeposited onto a cathode such as stainless steel which has an electrically conductive yet non-adherent surface, at the end of the process the deposited layer can be stripped off undamaged and form a solid metallic product in its own right. This technique is used for specialist technical applications, or for high value decorative work, because of the high amount of energy required. Products such as electric shaver foils are made in this way.

Another technology related to electrodeposition can be termed 'occlusion plating'. In this process, chemically inert powders are suspended in an electrolyte and are codeposited with the metal onto a cathode. The result is an adherent composite film. An example is offered by Ghouse (1980), who achieved mechanical strengthening of copper, without reduction of electrical conductivity, by dispersion of $6-22$ % of silicon carbide in the electrodeposited metal.

B.A.J. Vickers Ltd offer a wear-resistant coating service under the trade name 'Tribomet'. This involves occlusion plating chromium carbide particles into a cobalt matrix, and improves wear and abrasion resistance as well as resistance to corrosion and oxidation.

9.4.4 *Chemical deposition*

Long before electrolytic processes were understood, man had known chemical methods for producing a film of one metal on the surface of another. This is now known as immersion replacement, and will be familiar to anyone who has inserted a steel knife blade into a copper sulphate solution and seen a miraculous transformation of the surface colour. The process is self-limiting because deposition stops when all the substrate metal is covered; the very thin coatings so produced are usually of no value for technical purposes, as corrosion and wear resistance are too low because adhesion is poor, unless special complex solutions are employed.

More interesting is autocatalytic deposition, in which chemical reactions take place at a workpiece surface, with deposition of metal previously dissolved. As the metal deposit is itself catalytic, the process is continuous until stopped by removal of the workpiece from the solution. The process is commercially viable only for a limited number of metals, of which nickel is the most important, this always being co-deposited with up to 10% of phosphorus or boron because of the chemical constitution of the solutions used. Nickel can be deposited on steel, cast iron, aluminium and on most conductive and non-ferrous metals. The coating thickness is very uniform, even over the most intricate shapes, so that problems of indifferent throwing power of electrolytic nickel baths are overcome. These nickel deposits are very corrosion-resistant in many media. Occlusion plating can be carried out with this process for example, Montgomery Plating Co. Ltd offer a proprietary 'Niflor' process in which nickel and polytetrafluoroethylene particles are codeposited by the autocatalytic process to generate a surface with good corrosion resistance and also good bearing lubricity properties. Copper is another familiar material which is autocatalytically deposited, being particularly important in application to plastic printed circuit boards, although it is also applied to other plastics as a substrate for subsequent electroplating. Electroless processes have been reported for gold, rhodium, palladium, cobalt, silver and zinc. Recently, Chapman (1984) has reported the autocatalytic deposition of tin, but the process is currently too slow and of insufficient chemical stability to be commercially viable yet.

Although not strictly comparable with the previous processes, mechanical plating may be mentioned here because of the apparent similarity to the autocatalytic process insofar as a chemical 'soup' is used as an aid to production of a surface coating. The mechanical plating process applies a metallic surface coating to a metal substrate immersed in a liquid containing primers, chemicals, soft metal flakes and glass balls which, when tumbled together in a barrel, impact on the substrate surface and weld the suspended metal flakes to it. Brooks (1983) shows how malleable cadmium, zinc and tin can be deposited onto ferrous parts without risk of hydrogen dissolution and subsequent metal embrittlement, and how copper alloy parts can be coated in the same way. Thickness of deposits is from five micrometres up to those equivalent to hot dip galvanising. The process is economically attractive compared with electroplating, but of course is limited by the geometry of parts which can be treated, and by the limited variety of soft metals which can be deposited.

9.4.5   Conversion coating

In the technology of conversion coating, a substrate metal which needs protecting from corrosive influences is modified on the surface to produce a chemically resistant compound which is itself virtually a corrosion product. These conversion coatings alone may be sufficient in some cases; in others they are prepared as a good basis on which paint coatings can be applied, and to take care of occasions when chipping occurs, as in automotive practice.

Three main methods are used to convert metal surfaces into these chemical coatings which, in order to function properly, have to bond very well to the basis metal without a significant volume change which could cause cracking or peeling of the coating.

Phosphating is very commonly applied to steel, which is treated with phosphoric acid mixtures of accelerators, oxidant and dissolved metals, to produce an iron phosphate coating. Zinc and manganese phosphate coatings are used as undercoats and as oil carriers. Phosphating is also applicable to zinc, cadmium and aluminium.

Chromate conversion involves treating the material with an oxidising solution containing chromium and other metal ions. Steel substrates are normally electroplated with zinc or cadmium before the chromating process which, however, can be applied directly to aluminium and its alloys, zinc, cadmium, magnesium and tin.

The third main process for chemical conversion is anodising, which is particularly applied to aluminium, although it has been reported as applicable to pure iron. When anodising, a sulphuric acid electrolyte is normally used so that the correct crystal structure of hydrated aluminium oxide forms upon the substrate surface, to produce a porous but adherent coat. This coat is relatively weak when first formed but is strengthened by recrystallising processes such as treatment with hot water, and can be made to absorb dye-stuffs to provide a protective and yet decorative effect.

It is not appropriate to provide any further details of the operation of these processes, except to point out that they are often very dependent upon the type of alloy being worked upon, and their production should be left to experts.

9.4.6 *Vapour deposition processes*

Several techniques exist for coating a substrate surface or for modifying it to improve the performance in specific service conditions. Four main methods, taken from the wide gamut available at the present time, will briefly be discussed for illustrative purposes only.

Vacuum evaporation of a metal onto a substrate is an old and familiar technology. Essentially, any substrate material which does not produce too much gas can be worked upon, including metals, glasses, ceramics, plastics and paper. According to the design of the system, virtually any metal can be deposited as a coating, as can some non-metals such as silicon monoxide, silicon dioxide, cadmium sulphide, magnesium fluoride and many others. The coating material is heated electrically in a vacuum chamber and evaporates to condense on all cool surfaces in a line of sight from the original source. This means that the workpiece is coated, but so also is any part of the equipment not shadowed by workpieces. Because of the need for a vacuum chamber, in general only small objects are processed, up to, for example, the size of head-lamp reflectors for automotive use. The film is always very thin so that the process is more suitable for decoration and reflective requirements, rather than for production of wear- or corrosion-resistant coatings for engineering applications. Grace (1984) offers a recent general article on vacuum metallisation techniques and the types of plastic substrate which are best suited as workpieces. Of particular interest are his design pointers, which illustrate the forms of material section that can easily be metallised and those that cannot.

Sputtering techniques have been available for many years. These also depend on a low-pressure enclosed system and operate by ion bombardment of a target to produce at least one of the depositing species. Until very recently, this process has tended to work at a slower rate than direct evaporation, but it is suitable for specialist coatings, particularly for the production plating of machine tools.

Matthews (1984) discusses ionisation-assisted reactive evaporation, in which a vapourised metal reacts with a gas such as nitrogen, oxygen or acetylene to produce a nitride, oxide or carbide deposited on a workpiece heated to above 400 °C before coating. The process essentially injects reactive ions into the substrate surface, converting it chemically to form a very adherent layer. El-Sherbiny & Salem (1981) discuss the general technique of ion plating to protect surfaces in the aerospace industry, using nitrogen in particular, while Anon.(1982) discusses a nitriding process developed in Bulgaria, claimed to optimise nitrogen diffusion to saturation in a heated surface layer of steel. This ionic deposition process is claimed to be five times faster than is the rate of conventional nitriding, offering reduced thermal deformation of the workpiece. Ion plating has also been used to prepare iron–chromium alloys with good corrosion resistance against sodium chloride solutions, made in that amorphous (non-crystalline) state of metals which is of such absorbing interest in this decade. Diegle (1981) offers brief details.

A related method, offered by Robert Stewart (IVD) Ltd in England is the 'Ivadizing' process, in which ions of evaporated aluminium deposit on a substrate to form a soft ductile coating − a coating with properties nearly identical to those of pure aluminium which can be further protected with a supplementary conventional chromate conversion treatment.

Chemical vapour plating, unlike the previous processes, does not depend upon a vacuum system but operates by thermal decomposition or reduction of gases containing those elements which are to be deposited upon the heated substrate surface. A familiar early example is the decomposition of nickel carbonyl gas to produce pure nickel on surfaces, but many other materials can now be deposited by modern techniques. Typical applications include deposition of oxidation-resistant coatings such as silicides, or wear-resistant coatings such as carbides or borides. A potentially severe drawback is that the substrate often needs to be heated to several hundred degrees to allow deposition to occur, which might adversely affect the metallurgical condition of some alloys.

Glow discharge techniques involve polymerisation of organic monomers in a gas discharge to deposit a polymer film over the total surface of a workpiece. It offers the production of pinhole-free organic films with no volatile components in them, and most organic monomers can be polymerised. Typical applications include coatings for steel, coatings on leather and fabrics, but particularly the insulation of films in the electronics industry as, for example, when polyparaxylylene coatings are deposited on printed circuit boards by what is, generally speaking, a low-pressure and relatively inexpensive process.

9.4.7 Hot dip and diffusion processes

The self-explanatory term hot-dip implies immersion of a workpiece in a molten coating metal, so that interfacial alloying will occur. After removal from the metal bath the coating solidifies and provides a good metallurgical bond. The product, however, is not geometrically so uniform as is the case with electroplated coatings, but tends to be more reliable for use in severe environments. Metals commonly coated onto substrates in this way include aluminium, lead, tin, lead–tin solder, aluminium–zinc alloy and zinc; each of these techniques has its specific process method with optimum immersion times and temperatures, and its unique and well-defined range of dimensional and physical parameters. Probably, immersion in molten zinc is the most familiar process under the generic name 'galvanising', in which graded composition zinc–iron alloys are produced on an iron workpiece surface, the surface being of course zinc-rich. In addition to providing a hard and wear-resistant coating, the overlying pure zinc is available for sacrificial use in the event that underlying iron is exposed to corrosive environments. The life of a galvanised steel part varies of course with the severity of exposure and the thickness of the coating. As an example, 100 micrometres coating thickness in an outdoor, non-polluted rural environment should last almost 35 years before needing maintenance treatment, which can often be as simple as applying a coat of paint. Galvanised steel can pose problems if welding has to be carried out after the galvanising process, particularly likely when large structures are to be assembled. Relevant details can be obtained from the Zinc Development Association in London, England, but, as with all these coatings, any abnormal assembly operations should be discussed beforehand with experts.

Diffusion coatings are made by taking a metallic workpiece and diffusing into its surface a different metal, which may generate an alloyed or interstitial coating on the surface. The ingoing metal then diffuses towards the centre of the workpiece in accordance with normal rate-controlling diffusion laws. So the outer layer may comprise a pure or highly concentrated coating metal but with a composition becoming progressively more dilute towards the centre of the workpiece, but always with very good bonding. Very often a cementation process is used, in which inert carrier powder such as sand, and a metal such as aluminium, chromium, boron or zinc, are mixed together, packed round the workpiece and then subjected to a high temperature. In other variants, a halogen compound is used as a carrier to transport diffusant metal to the workpiece, being itself recycled during the process. Deposition of aluminium, nickel and cobalt alloys, of chromium and of silicon on steels, are reasonably well-known. Mention must be made of the sherardizing process, the diffusion into steel of solid zinc (as powder) as an alternative to the galvanising process. Recently, Chapman (1984) has discussed preparation of copper−tin alloys by thermal diffusion. A general feature of all these diffusion coatings is their suitability for bulk processing of small parts, while providing very uniform layers over the whole exterior surface.

9.5   Ceramic finishes

Included in this section are deposited ceramics and also vitreous enamels. Although both of their technologies are quite complex, they are not so confusing to the layman as are metal finishes; consequently they will be discussed much more briefly.

The flame-spraying process of deposition has already been described, and can be used to deposit non-metallic coatings, of which alumina, zirconia and tungsten carbide are typical. These coatings generally adhere well to a suitably prepared substrate, and typically are used to increase the working temperature of those devices planned to live a short but strenuous life, for example rocket nose cones and motor parts.

An alternative process is a modification of the chemical vapour deposition method, which yields surface coatings with extremely fine and controllable grain size, and thus with very high resistance to friction, corrosive wear and extreme temperatures. These coatings are used to improve the life of machine tools; deposition of silicon carbide on to metals, ceramics or graphite are typical applications.

Vitreous enamels have been used in consumer products for half a century or more, and comprise a ceramic coating applied usually by a wet process to a metal surface and then fired to produce a fused, glassy layer. Real vitreous enamel is usually based upon glass fused onto steel or cast iron at temperatures around 800 °C; the oft-confused term 'stove enamel' refers to baked organic coatings applied and cured at much lower temperatures. A good introduction to vitreous enamels is offered by Clarke (1981). Although vitreous enamel in its basic state of slip (glass frit with a liquid medium) is a liquid, other application techniques are available, for example deposition by electrostatic spraying, as indicated by Vaughan (1981).

However applied, the substrate must be adequately prepared both by chemical cleaning and by mechanical blasting to develop a rough surface which will provide a good key for the coating. Not all steels are suitable for enamelling, so a specialist enameller should be consulted before specific steel types are decided upon. Although the substrates commonly used are cast iron, steel sheet, aluminium alloys and copper, nevertheless bronze, gold and silver are also enamelled for specialist and decorative work, with the proviso that the temperature needed to fuse the glass frit must be less than the melting point of the metal or alloy. In general, finished vitreous enamelling is very resistant to most exterior influences, such as atmospheric and chemical corrosion, as well as to oxidation. The enamels offer a very wide colour range and considerable thermal shock resistance, provided that the thermal expansion values of finish and substrate have been matched. Enamels are fairly hard and generally quite scratch-resistant, but will not stand heavy deformation of a substrate, as they then tend to flake off. A familiar example of vitreous enamelled finishes is the humble kitchen saucepan.

9.6 Organic (paint) finishes

Paints have been used by mankind since time immemorial, and thus the subject of painting has become technologically very complex. On the other hand, most practical people have some idea of the variety of paints which are available, and of at least some of the application methods, several of which have briefly been described in Table 9.4. It should be sufficient, therefore, merely to remind the reader of some key points about paints, and about organic finishes, bearing in mind that all paints are not organic, neither are organic finishes all necessarily paints.

Taking paint first, the two primary functions are decoration and protection of the substrate. Paints are based upon a number of basic polymers which include alkyds, vinyls, polyurethanes, acrylics and epoxies. This is no place to enter into even a brief description of the chemistry of these or of the many other types, but the short list of types is offered as a reminder of the complexity of the subject, and to make the reader realise that he is unlikely to know instinctively how to make correct decisions when specifying paints for his application. Paints are based on a mixture of pigment with a liquid vehicle and with other added materials such as binders, cure accelerators, solvents, viscosity modifiers and many more. The pigments are included to impart colour, to provide body, to absorb ultra-violet light and thus to protect the rest of the paint components and the underlying substrate, and to act as a corrosion inhibitor. Pigments may be synthetic organics or they may be mineral-based, such as oxides of titanium, zinc or lead; they may also be metallic as in the case of aluminium or zinc-based paints for corrosion protection. Other inorganic or mineral compounds are used as paint pigments and for protective purposes; calcium plumbate is an example, providing a yellow colour and at the same time good additional protection for hot-dipped galvanised steels for which it forms an excellent primer.

Paint as a bulk liquid is of no use in an engineering product. It can only operate when spread uniformly and (usually) thinly as a coherent film which has dried and hardened. The drying may take place through simple loss of solvent by evaporation, but in the more durable paints occurs by oxidative reaction of components of the paint, leading to formation of insoluble organic films. The viscosity of the paint must be held within strict limits during application and during drying, so that run-off and sagging are avoided, leading to the need for inclusion of even more additives. The curing process by which the film ultimately hardens may operate at room temperature, but more often is carried out at higher temperatures, as is most usual in industrial applications.

An industrial finish differs from a decorative finish through the emphasis on its function, since protection must be the main requirement and appearance is secondary. Such paints are used for new products and also for industrial maintenance, being used to protect equipment or structures from deterioration and corrosion by industrial or other atmospheres. Industrial maintenance paints are regularly applied to structures such as the Forth Bridge. Taking this as an example, it can be seen that formulation of the paints must be carried out carefully, to provide as much application latitude as is possible; the situation of the painter can never properly be controlled or often even foreseen in terms of ambient temperature or humidity.

A key factor in specifying a paint finish is adequate durability. For short-lived consumer goods there is obviously wide scope for selecting paints, since most of them will last in service under any conditions for a short period. Where more durable coatings are needed, then it is possible that a paint film applied as a liquid may not be quite thick enough, and other materials and methods of application are called for. Protective coatings based on polymers provide a coating thickness greater than that derived from conventional liquid paints, and comprise solutions or gels of polymers in solvents, which are then totally or partly dried; an example is PVC plastisols, used as insulation on electro-technical products.

Plastics themselves comprise an increasingly important class of substrate, for which the applied paints must be formulated specially. This is because paint vehicles must be compatible with the substrate plastic, of which some classes are easily cracked by exposure to solvents. Although plastics can be inherently self-coloured with pigments or dyes, they are often painted for economic reasons when a small batch of mouldings must be produced in a specific colour, or when surface defects would otherwise be apparent. There is also a need to use compatible painting systems for plastics and metals together in the modern automotive industry. Increasingly this compatibility is being extended to the state where whole cars comprising metal and plastic components are painted and stored as an entity. The main plastics to be treated in this way are pretreated polypropylene, polypropylene-based copolymers and elastomers, polyamides, acrylonitrile−butadiene−styrene, polyesters and epoxies. In all such cases, the curing temperature of the paint must not exceed that at which the base polymer loses dimensional stability.

Plastic powder coatings based either on thermoplastic or thermosetting resins are increasingly being specified for decorative and protective applications. Thermoplastics widely used include polyvinyl chloride as a powder or as a plastisol (a gel-like solution in a high-boiling plasticiser), polyurethanes, some polyamides, cellulose based compounds, polyethylene and fluoroplastics. They are typically used for coating or lining industrial plant, marine components, wire-work as in the case of dish-washer grids, rain-water goods, window-frames, electrical insulation and many other applications. The literature is replete with specific information, but mention will only be made of a paper by White (1984), who discusses the economical use of fluoroplastics for coating metal surfaces to improve wear resistance and surface lubricity. White points out that these polymers offer stability of mechanical and electrical properties over a temperature range of about 500 °C, and can be made to adhere adequately on a steel substrate at thicknesses of up to 30 micrometres, provided the appropriate production process is chosen. This can involve spray coating or a dip−spin coating in the case of small components.

Thermosetting powder coatings, on the other hand, are quite different. These materials, based upon polyesters, epoxies, polyurethanes or acrylics, are deposited on the cold workpiece as dry powders, generally by electrostatic spraying, although fluidised bed coatings can be applied to preheated components. In any event, the substrate is heated to melt and thus cross-link the thermoset material, which provides an average coating thickness of 50−60 micrometres, twice that from a normal liquid paint film. These materials can be thought of as vitreous organic enamels applied at a low temperature, but without the upper temperature service performance, and without quite the same high level of environmental resistance. However, for electrical components, pipe-linings and for many other uses, they provide a technically adequate and economically interesting solution to the problem of protecting products.

10

Materials reliability and service life

For any company and its products to be successful in the market-place, most products must have a worthwhile lifetime and reliability in service to match the expectations of the purchaser. In order to achieve this the designer, among others, must have some feel for the ways in which the useful life in service of a material can be estimated, and must reflect these factors in his design. The purpose of this chapter is to discuss briefly some aspects of prediction of service life and the design needed to ensure that this life and a safe product results. Consideration of the design decisions for correct materials choice and the mechanisms of failure of materials in service is focused on metals and plastics as examples of the whole field of materials.

10.1.   Predicting service life

Materials respond to their environments in a time-dependent manner, which is usually forecast from accelerated laboratory experiments; results of experiments are compared with what has been seen under real service conditions to prove and extend their validity. Provided that accelerating factors for laboratory tests are correctly chosen, and that the tests themselves are correctly representative of real life service, it is reasonable to expect that results can be used to calculate what must be done to achieve a predetermined life in service. Laboratory tests which attempt to simulate, in an accelerated way, real life exposures must of course be truly representative in a mechanistic way. For example, when testing a metal for corrosion in a liquid medium, it is necessary to know whether in real life air and other materials will be present at the metal surface in addition to the corrosive liquid: laboratory tests in which a metal is totally immersed in a corrosive medium will not provide responses simulating real life conditions in which the metal is wetted with the liquid for only part of the time.

Watson (1979) points out that the vast majority of failures in the engineering industry are caused by fatigue of metals due to an alternating stress being applied to the material. The alternating stress may vary cyclically or may change in some other way during service life, leading to what is known as spectrum loading. It is seldom practicable to undertake true spectrum loading tests in the laboratory, and herein lies one of the first hurdles which the designer must overcome in discussing this aspect of design with his expert consultants. Fatigue testing in the laboratory is most often determined by using a constant stress amplitude, giving rise to the familiar $S-N$ curve.

Static testing of materials does not yield a very useful view of the potential durability of a product. Separate components of a complex product may, however, be tested with simple or complex loadings, which are particularly useful for testing welded joints. Watson's conclusion is illuminating because he claims that efficient design with expensive starting materials can be made to produce the cheapest components, and the selection of material must depend upon an exact knowledge of service loads and materials properties.

Customer satisfaction with a product depends not only on its reliability, that is, the probability that it will fulfil its intended function for a specified interval of time, but on maintainability as well. Maintainability ensures that, in the event of failure, a predetermined and hopefully short interval of down-time elapses. In any system, maintainability and reliability are related in a rather complex manner, well described by Lamberson (1981).

Use of modern knowledge of materials performance in simulated service conditions can usually avoid the older problem of overdesign to ensure a permanently safe, if uneconomic, life, and frequently enables design to be such that it meets a specified and well-defined service life. However, it must be remembered that at the end of that controlled service life, the product must not be expected still to perform in a satisfactory manner. In other words, the product must progressively age and fail catastrophically, like a lamp bulb, not gradually as in the case of some British motorways and most people.

As early as possible in the development process, the designer must be aware of the need to design for safe operation throughout the projected service life conditions. Apart from the normal application of good sense, many standards now operate nationally to control requirements which must be complied with before a product can be marketed. During the design of a product, two types of service condition need to be borne in mind. The first is the normal working condition likely to be met during the whole of the intended service life, while the second is an overloading limit, which the product may be called upon to reach only on exceptional occasions and which, if repeated too frequently, would cause lasting damage or premature failure. It is the responsibility of laboratory test results to indicate the extent of abnormal overload conditions which can be borne.

Safe-life design was introduced immediately after the Second World War, when failure became recognised as a major factor to be considered in the embryonic civil aircraft industry. This criterion applies to components which may embody a hitherto undetected crack or defect which could lead to a catastrophic structural failure; a life limitation is accordingly imposed which reflects the maximum extent to which the crack might expect to grow. In this way, significant fatigue damage could not develop during the permitted service life of the aircraft or other product. Unfortunately this system demands the premature scrapping of structures which do not fail in the prescribed lifetime.

This philosophy was succeeded by the fail-safe criterion, applied to structures where there is sufficient inherent strength such that any fatigue cracks or other problems which arise in service would not lead to catastrophic failure before the next routine service inspection. The onus here is on the inspection to find all likely hazards, and to estimate whether a fault found is likely to grow into a dangerous one before the next routine inspection takes place. The onus too is on having an absolutely regular inpsection schedule which is never interrupted.

Fig. 10.1. Stages in designing for safety. (Copyright: 1985 Standard Telephones and Cables plc.)

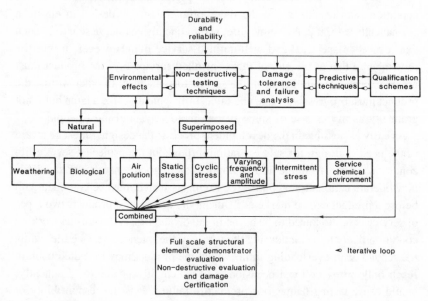

An even later philosophy is that of damage-tolerant structures, in which the original material is assumed to embody flaws which might ultimately lead to cracks and crack growth, the larger ones being designed out of the material by efficient processing methods so that those remaining are below a critical size. The mathematics of critical crack size and failure are well known and are available in many standard texts, of which that of Collins (1981) is typical. This philosophy depends upon superb process control and inspection at the manufacturing stage.

The discrete stages illustrated in Figure 10.1, in the process of designing for safety, involve people other than the designer who must have access to basic data on materials, to permit property comparisons and predictions. The methods of building adequate data bases are discussed by Haugen (1982*b*).

Many standards operate in the UK and in other countries to govern the production and operation of products from a safety viewpoint. Table 10.1 identifies just a few of those considered most applicable to general engineering products.

Table 10.1. *Some standards for product safety*

International Electrotechnical Commission

IEC 227/245	Insulated flexible cords
IEC 335	Household electrical appliances
IEC 553	Non-metallic items − fire resistance

British Standards

BS 415	Mains-operated household sound/vision equipment
BS 3456	Household electrical appliances
BS 3861	Office machines
BS 4644	Data processing equipment
BS CP95	Fire protection for EDP installations
BS 4086	Maximum surface temperature of heated domestic equipment
BS 5760	Reliability of systems, equipments and components

Underwriters Laboratories Standards (USA)

UL 726	Boiler assemblies − oil fired
UL 353	Limit controls
UL 430	Electric waste disposers
UL 82	Electric gardening appliances
UL 972	Burglary resisting glazing materials
UL 746D	Polymeric materials − fabricated parts

10.2   Design decisions

A preliminary and most important design decision in any product development is correct choice of the material. Much has been said in earlier chapters about the characteristics of materials and their property limitations. The choice of material must depend upon correct specifications being available at the right times. Waterman (1979) defines market specifications as identifying the need for the product and also defining the functions it must perform within the expected service environment. These should then be followed by design specifications, which embrace component parts, dimensions, materials and the rest. The product process specification shows how the product will be made; this aspect of the designer's interest is addressed in Chapter 12. Finally, a quality control specification is needed which confirms that the other specifications have been met. Waterman provides some enlightening illustrations of failed products which were not properly specified originally. One was caused by the impact anisotropy of high impact polystyrene resulting from orientation of the material during flow in an injection moulding process.

Failures in a product have an effect on the profitability of the enterprise producing it. Haugen (1982c) illustrates a profit−loss estimate for the production loss of parts, plotted as a function of the probability of failure. Expected profits decrease as the probability of failure increases, because of an increase in the cost of replacing or repairing parts. Too low a reliability level could result in a total loss of business. A much more comprehensive treatment of the narrower field of polymer composite reliability is offered by Kaelble (1983) in a CAD/CAM handbook.

Perhaps the most important single problem when selecting materials for known service life is the problem of fatigue. Normal fatigue is a progressive phenomenon, brought about by cyclic application of strains and ultimately leading to cracking and failure. Even in a 'static' part not subject to significant mechanical vibration at those amplitudes and stresses likely to cause fatigue, low cycle fatigue can occur merely as a result of the cyclic thermal history during product life. If a machine part undergoes cyclic temperature changes, and the thermal movement so induced is constrained, even partially, cyclic stresses and strains will occur which can lead to this type of fatigue damage. Test results on metals have shown that thermal cycling can be an even more likely cause of failure than mechanical cycling regimes at higher temperatures. This should remind the reader that another critical parameter in materials selection is the operating temperature of the product. An innovative package is now available, Anon. (1983c), in which a chart is provided showing design constraints upon plastics operating at high or low service temperature.

10.3 The lessons of failure in service

Most practising designers and engineers can point to material failures in service, but perhaps are not careful to review the reasons for failure. The effect of orientation in injection moulded polystyrene on the impact resistance has already been mentioned; other examples appear in the literature. Some examples, also from Waterman, include the lack of wetting-out of reinforcing glass fibres by resin-filler in sheet moulding compounds (leading to resin-starved areas of low mechanical strength) and the incidence of porosity in zinc die castings, weakening the structure at key points. Harrison (1980) presents details of 100 failures since 1954 in metal welding. Lessons are drawn about those errors of design which can lead to collapse of structures.

Anticipating the next section, the possibility of failures at interfaces between materials where electrolytic action and the corrosive effect of water and oxygen can cause problems, and acknowledging the difficulties of bonding dissimilar

Table 10.2. *Some causes of material and product failure*

Direct effects
       Elastic deformation
       Yielding
       Ductile rupture
       Brittle fracture
       Impact damage
       Fatigue
       Fretting
       Creep
       Corrosion and stress corrosion
       Wear
       Thermal cycling
       Galling and seizure
       Spalling
       Buckling

Failure-inducing agents
       Force
       Time
       Temperature
       Reactive environment

---

materials with significant thermal expansion differences, we can pass on to Table 10.2 adapting from Collins (1981) some of the more common failure modes in materials and products which have been observed in service.

10.4   Failure of materials

A major source of failure of metals in service is corrosion, often caused by interaction of an electrically conductive liquid with a pair of metals which, between them, constitute an electrolytic cell. Corrosion of this type can occur within a single metal which has inclusions of alloying elements or impurities in it, and is often seen in aluminium alloys. Corrosion may arise through several different mechanisms including direct chemical attack, corrosion fatigue, crevice corrosion, dezincification (in the case of brasses), erosion corrosion, fretting corrosion, graphitisation, hydrogen embrittlement, intergranular corrosion and stress corrosion. Table 9.1 provides a galvanic series of metals identifying those combinations which, through being most widely spaced in the list, are the most likely to be corroded even by mild electrolytes. There is, however, no absolute list order from which unshakable predictions can be made about which metal in a combination will suffer the most corrosion. The relative positions of metals in the Table 9.1 sequence depend upon environmental parameters including temperature, oxygen concentration and the concentration of some dissolved ions such as chlorides and sulphates.

The term 'corrosion' is seldom considered in the context of ceramic materials, which are renowned for being very resistant towards gases and liquids. However, there are cases where they will degrade in service. For example, although many ceramics are based on oxide compounds, their use in a high temperature oxidising atmosphere with a high sulphur content can lead to the formation of sulphates and surface degradation. Some carbide ceramics are oxidised at very high temperatures, as is silicon carbide, but boron carbide and graphite cannot be used in oxidising atmospheres above a moderate 500 °C. Corrosion by liquid metal contact occurs at high temperatures with certain specific combinations, including silicon nitride which must not be allowed to contact molten alloys containing copper, and fused salt and molten glass which will attack some ceramic materials. It is assumed, however, that the present reader is not concerned with these extremes of performance in ceramics, but more with the benign environments in which most engineering products have to perform. Although ceramics are mostly resistant to aqueous solutions, except for very strongly alkaline ones, it must be remembered that those based on silica will react with hydrofluoric acid, and some glasses, particularly soda glasses, are etched by alkalies and by some strong acids.

Fatigue has been mentioned before. Details of the exact nature of fatigue failure and how to avoid it will not be discussed here, as many excellent texts are already available. In fatigue failure, cracks, if not already present in the material, are first initiated and then propagate to an unstable size through operation of applied nodes, which may arise from mechanical factors or from temperature differences. Corrosive environment will often exacerbate fatigue failure but such a simple additive as pure water can effectively reduce the strength of structural steels.

Although an obvious palliative against fatigue from cracks is to use material which has no cracks, this is rather difficult in some materials unless a 100% effective inspection is carried out, and this is not usually economic. However, the designer can play his part in reducing fatigue failures by avoiding, for example, sharp fillets and other stress-raising configurations in the design, and by paying due attention to the methods of manufacture which will be adopted. Basic design calculations should be made on the realistically estimated stresses which are likely to be imposed upon the whole part in service, rather than on the concept of an average stress on an average section. Mechanical fastening of parts will locally increase stresses through the clamping force involved. Provision of multiple loading paths and symmetrical placing of weight or force in a loading sequence will reduce localised stress build-up, although if these are unavoidable geometrically, the joints or sections can always be reinforced. As joints are main regions for fatigue crack initiation, jointing methods to be used should come under design scrutiny and not simply be left as the responsibility of an engineer further down the manufacturing line. Due note should be taken of the material's characteristics in its fatigue – stress cycle, as illustrated in British Standard 3518, so that permissible stress amplitudes and mean stresses are not exceeded. It is claimed that fatigue strength of welding joints can be improved by a number of methods, as shown by Baxter & Booth (1979). Peening and grinding of the weld toe to remove slag, and subsequent remelting by the TIG process, produce a smooth transition between welded metal and the parent plate, increasing the strength significantly. Autogenous remelting of the weld toe using an arc plasma welding process seems a reasonable method of increasing fatigue strength by eliminating some of the high- skill content needed. In the context of reviewing defects in welds, these should be considered essentialy as small castings, with the difference that the fused weld metal constitutes the weld itself, surrounded by a welded substrate zone affected by heat from the process itself. This configuration can lead to a variety of defects including microcracking, oxidation, porosity, absorption-desorption of hydrogen, inclusions of slag and the effects of low ductility in the weld metal.

Nothing has been said, so far, about creep. This is the time-dependent strain which occurs under continuous loading of a material. Depending upon the material, creep may be manifest by a change of shape, particularly in the case of plastics, and it may ultimately lead to fracture with little sharp distortion. In self-loading devices such as some fasteners, creep compensates for applied stress, and leads to relaxation and hence a more stable long term situation. At low temperatures, the creep strain varies exponentially with time, and an increase in temperature or stress increases the creep rate.

At high temperatures (those greater than half that of the metal's melting point in absolute temperature) there are three well-defined stages of creep; in the first the rate decreases with time; in the second the rate becomes constant; while, finally, the rate increases until the material ultimately fails. Although accelerated creep test methods exist, unfortunately, they do not often properly simulate those metallurgical changes occurring in the materials in real service: these changes alter the creep properties observed. Accordingly, the designer is advised to be very conservative in his estimations of creep of metals, even though test methods are reflected in prestigious documents such as British Standard 3500 and ASTM E139.

For plastics, creep also depends on the applied stress level, the temperature and its duration, and also upon environmental factors such as humidity concentration and previous processing history. Some plastics exhibit viscoelastic creep behaviour demonstrable even at ambient temperatures. This phenomenon implies that some recovery occurs once the applied load is removed, and in a few cases may be complete. Because the subject of creep test in plastics is in its infancy compared with metals, it is often desirable to circumvent its likely effects by conservative design to reduce stress levels to safe values.

10.5   Product quality

'Quality' of a product strictly means that it conforms to the expected performance requirements. It does not necessarily mean that the product is very good value for money, or that it provides many more facilities and functions than would be expected in a product of that price. 'Quality' means 'conformance' and, accordingly, quality control of a product during and after manufacture, and quality auditing of the project which gave it birth, are strictly about checking against predetermined standards.

Quality control embraces testing and measuring performance with the aim that the component will be accepted, rejected or returned for rework. Quality control may involve 100 % inspection of every item in a batch, or the inspection of random samples taken from a batch. The number of samples to be taken for quality purposes, from a batch of a given size, is predetermined in available tables. Owing to the problems of human error and the failings of inspection equipment, 100 % inspection of a batch does not automatically mean that all defective parts will be completely eliminated. Furthermore, evaluation of a component performance by measuring a single datum point is not necessarily a valid procedure, as single values are not usually suitable for prediction of service performance, a set of replicate results being spread in the normal Gaussian manner. Admittedly, the spread may be quite small, but it is there.

Quality auditing of projects involves more than selection of materials for production. It involves the determination of the status of specifications which control the activity, the availability and implementation of a quality plan, and the operation of progress reviews in which the designer may play a key part, at least initially. It depends on the availability of documentation and records, fabrication control and is even concerned with the status of measuring instruments, the testing and inspection of deliverable items and eventually product qualification against an imposed performance specification.

Also included under the heading of 'Quality' is analysis for product safety. This may be carried out simply by applying a few functional tests, or a formal procedure can be invoked in which all the potential hazards of the product in normal and fault operation conditions are identified and ranked. For example, in an electrical product one may consider that the main hazards to the user are electric shock, fire, explosion or implosion, liberation of toxic gases, generation of harmful radiation, emission of particulate materials harmful to health and the possibility of the product with sharp edges or of great mass cutting or crushing someone nearby. These hazards can be ranked in, say, five levels, in which rank 1 is a very low occurrence rate (perhaps one million to one) up to rank 5 with an occurrence rate say at more than ten to one. This system can rate a product in the form of a pictorial safety profile, in which the hazard occurrence ranking is plotted against a criticality ranking. In this representation, an area with low criticality and low hazard occurrence would represent safe operation, whereas the existence of an area at the other extreme would require that the manufacturer undertakes some action to make his product safer.

Many methods of assessing quality of materials are, of course, available in the literature − just three will be mentioned. A low frequency eddy current system can evaluate structural changes in cast iron. Beard (1980) suggests a system that will aid identification of the difference between one casting and another standard one known to be good. A variable magnetic field is applied to the piece part and cathode ray traces are thrown up to indicate differences between good and defective parts.

Intergranular failures in metals range from temper brittleness and stress corrosion cracking to reheat cracking and creep embrittlement, and might be associated in metals with the presence of specific impurity elements. A different impurity hierarchy accounts for each of these failure problems, but all can be appraised in terms of a unified treatment. Seah, Lea & Hondras (1981) propose an overall assessment parameter, the fragility index, which they developed. This index may be used in alloy design to predict in a quantitative manner the effects of improved control of impurity tramp elements.

Moulders of plastic articles need quick and simple tests to determine whether their parts are being made to specification in real time on the shop floor. Moritz (1980) suggests a regime of tests involving accurate measurement of part weight, measurement of strain by using a hand-held polaroscope, and instrumental determination of moisture and measurement of flow into the mould by using a flow tab taken off the injection runner. Such methods could be used as a quick form of quality control on the shop floor and would thus reduce the load in the inspection department later down the line.

10.6   Use of statistics

The concepts of materials reliability, service life, variability and effects of more than two parameters operating simultaneously on a material, inevitably lead to the need for statistical calculations. This text is not the place for any detail, but some adequate books are listed in the bibliography. Haugen (1982*d*) has prepared a series of articles on the theme of modern statistical materials selection. As materials specification becomes ever more critical, there must be a parallel increase in reliability to reduce those uncertainties still remaining. With the concept of random variables, materials properties can be described to reflect the spectrum of possible values that they display in the real world. Classical equations on materials behaviour can be adapted to yield useful results in terms of the probability of failure or, alternatively, the minimum allowable reliability. An example; although it is usual to aver: 'this unit has a safety factor of 2.3', it is much more accurate and informative to be able to state: 'the product made from this material has a 10^{-5} probability of failure after 200 000 cycles of operation.' Haugen (1982*e*) discusses statistical concepts such as random variables, probability and reliability, and presents statistical concepts to verify the relationship between measured random variables and the degree of that relationship. This can be done with easily available computer programmes, a simple example of which is solving the familiar problem of drawing the best straight line through a set of experimental points by the method of least squares.

11

Factors controlling the selection of substitute materials

The Concise Oxford English Dictionary defines 'substitution' as 'a person or thing performing some function instead of another', or alternatively 'putting a person or thing in exchange for another'. In the context of tangible products, substitution of materials, therefore, implies that a change can be made to one or more materials previously selected for use in a product.

11.1 Materials are not totally interchangeable

A moment's thought will show that in no case can any one material fully and exactly substitute for another, since if all their properties were congruent in all respects, they would constitute the same material by all the normal definitions of materials science. Chapter 3.2 has already explored the concept of envelopes of properties, and Figure 11.1 illustrates how two materials may have envelopes of properties sufficiently similar that a product in which the substitution is made can essentially be the same as the original, with virtually the same performance parameters. As an envelope of properties is a mix of mechanical, physical, thermal, electrical, chemical and other properties, it is self-evident that in one of the materials of Figure 11.1, there might be a preponderance of, say, particularly good electrical or thermal properties, against which the second material would be somewhat deficient, but which may in turn offer better mechanical properties. Provided that the material originally used in the product was sympathetically chosen with a sufficient (if accidental) margin of overdesign, then reductions or increases of certain performance parameters will have no significant effect in service. For example, if in a coffee mill the outer casing is, by a substitution, made twice as strong in tension as the original was, there will be little effect on the operation of the mill or on the user's concept of its performance, even though the substitution may have been essential for the manufacturer's profitability.

A second moment's thought will show that if the substituted material differs too widely from the original, then the performance of the product may be modified so significantly that effectively it becomes a new product; in this event the original raison d'etre for the substitution is lost, even though the final result may open up new prospects in the market place.

11.2   Reasons for substitution

The reasons for an enterprise initiating a substitution activity are many and various, as illustrated in Table 11.1. Technical improvements in materials technology, in processing and in fabrication may offer new opportunities; new legislation on health and safety arising from new lobbies in the market place, such as concern for the environment, may demand product changes. A potent motive for changing a product is very often the sheer will of a company to survive economically be reducing the cost of its products and widening their acceptability, or through upgrading the company image. A further reason illustrated by Gray (1980) is concern about the availability of strategic metals within any one country. This is discussed at greater length later.

Some examples can be given of the key reasons for attempting substitution of materials in an established product.

Fig. 11.1. Materials 'envelopes of properties' and their commonality for materials substitution. (Copyright: 1984 Standard Telephones and Cables plc. By permission of the Institution of Production Engineers.)

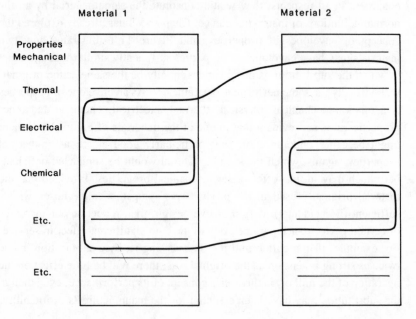

New production routes become available from time to time, either because technology advances and its improved methods of fabrication are publicised, or alternatively because a company itself may upgrade its production line with new plant and machinery, thus offering more opportunity to the product designer. Metals, plastics, cloth and even some foods are now machined or cut by the use of high-pressure water or high-pressure liquid metal jets. Up to 50 thicknesses of cloth suiting may be cut simultaneously by a water jet more cheaply than the older rolling knife methods. Plasmas are used to soften the surface of hard metals locally immediately before being machined, enabling conventional cutting tools to be used, while electron beam welding of metals overcomes many of the problems of screening a workpiece with inert gas blankets, as is required by certain other welding methods.

Easier manufacture generally equates to cheaper manufacture; this is sometimes simply effected by replacing materials in a product. Conversely, a new process may be adopted which may widen or, alternatively, restrict the range of materials which can be used. Recent changes with polymers involve, for example, altering curing agents used to cross-link elastomers. More active curing agents work faster, so that in an extrusion processing line the length of the heated curing section can be reduced, thus freeing valuable factory space.

In certain plastics, particularly ABS, snap-fitting of parts can be used instead of mechanical assembly by means of metal screws or other fasteners. This exploits the slight flexibility of some plastics and their elastic behaviour when snap-fitting is effected. This technique is now frequently used for holding telephone subscriber set housings together. For glass-reinforced plastics special chemicals are now available which begin curing the resin by exposure to ultraviolet light or even daylight. This initial curing hardens the material quickly and so shortens the time between laying-up and demoulding, and is thus a

Table 11.1. *Reasons for substituting materials in an existing product*

Reasons
New production routes become available
New materials become available
Reduction of cost to retain market position
Impact of new legislation
The impulse for novelty
Supply of key material is unreliable
Reducing dependence on imports
Upgrading of company image or of product range
Wider performance limits needed to satisfy new markets

particularly economical process for the manufacture of large items like boat hulls.

Made by the injection moulding process, plastic buttons for teleprinter and computer keyboards have for a long time been made by a two-shot moulding method, but a dyeing process has now been developed which offers a cheaper route for the product. Injection moulded television cabinets now largely replace earlier wooden models, reducing the costs of materials and of manufacture. Lastly, mention may be made of those injection moulded plastic screws used in some applications to replace rolled or turned metal screws.

Still on the subject of screws, and considering easier assembly of metals, several types of self-drilling metal screws are available, which drill their own pilot hole and then tap their fastening hole, thus eliminating a manufacturing process.

Substitution of cast or wrought metal alloys, and the drilling and other machining processes which they often demand, can now be made through use of powder metallurgy, by which metal powders are compacted and subsequently fired and sintered to near net shape. Hard steel moulds used by the plastics injection moulding industry can now be replaced in short production runs by softer moulds made of aluminium alloys which are very much easier to fabricate, if not so durable. Economically speaking, in the aircraft industry there is considerable pay-back in reducing the weight of airframes, and recent development of new aluminium-lithium alloys could see substantial replacement of normal aluminium alloys in the near future. For other substitutions intended to simplify and cheapen manufacturing processes, brief notice can be taken of the modern practice of staining and lacquering wood for furniture in place of labour-intensive and old-fashioned French polish, and also of the use of fasteners and glue to replace conventionally machined wooden joints such as dove-tails in drawers. Blister packs of card and thin plastic are now used for presenting many products in place of old-fashioned cardboard boxes, yielding economies both in material and in transportation and vending space. Metal assembly, in for example, the automotive industry, can be effected by adhesive bonding to replace or supplement spot-or fillet-welds. Lastly in this section on easier manufacture, consider the adoption of ice as a binder for moulding sand used in metal casting; replacement of the traditional resin binder simplifies reuse of the sand.

Official legislation and market place requirements imposed by consumer action are a strong force in achieving changes in materials chosen for products. Mention need only be made of recent health and safety requirements governing the past use, and future replacement of, asbestos by other materials in buildings, and the efforts to replace hexavalent electrolytes for chromium plating by trivalent ones to reduce mist hazards. In Sweden, however, an attempted complete ban some years ago on the use of cadmium in all its forms proved unsuccessful, largely because in some applications, such as surface finishing of running gear on vehicles and in some artists' colours, there is no real substitute. Nevertheless the action aroused public awareness of the cadmium toxicity problem, and substitution is being made wherever practicable.

The whole field of environmental pollution and effluent control has been under considerable scrutiny for several years now. In the process of zinc electroplating, for example, considerable efforts have been made to eliminate the traditional cyanide-based electrolytes and to replace them by acid solutions, solely to simplify effective effluent treatment. Successful efforts have achieved significant retardation of flammability of polymers by development of new additives and handling techniques — a subject already discussed in Section 8.5.

11.3 Product improvement

New materials, processes and techniques are widely publicised in the technical literature, and awareness of developments can often lead a company to improve its products, largely with the aim of entering the market-place with something novel. Some recent examples include development of plastic vehicle leaf springs made of carbon-reinforced epoxy resin to reduce unsprung weight, and the manufacture of melded plastic mesh, in which the cross-overs in what would normally be a woven fabric are instead firmly bonded together, making the material more suitable than conventionally woven fabrics for filtration applications. A domestic example of substitution was replacement of shellac gramophone records by successively improved variants of polyvinyl chloride, while a more down-to-earth development involved the successful replacement of leather shoe soles by a porous form of the same material. This particular development was a technical tour de force, but reminds us all that market research must be very thoroughly done; the development failed, despite the good performance of the shoe soles, because the buying habits of the ladies expected to purchase them dictate that shoes are rejected for fashion reasons and not because they have reached the end of their useful life; antique shoes are not acceptable on feet!

New technology leads the market to expect extra and different facilities and functions in well-established products. For example, liquid crystal displays have now largely replaced metal watch faces and form calculator displays as well. Heat-shrinkable plastic film displaces other packaging methods for covering white goods, palletised building bricks and other items, as a shipping protection and to permit some inspection before unwrapping. In the field of metals, the nickel—titanium memory metal trade named Nitinol is beginning to replace the bimetallic strip in thermostats, since unlike them it offers the additional function of transmitting significant force as well as movement. Super-plastic metals are being used for complex shapes at the wing roots in aircraft, and to replace plastics for the bezels surrounding cathode ray tubes in oscilloscopes. Another new materials replacement is that of copper in telecommunication cables by quartz-based optical fibres, although it cannot be pretended that the messages are transmitted by similar physical means. However, the end result to the customer is improved technical performance at lower cost.

Attention to materials substitution is often forced on a company needing to reduce product prices, or wishing to advertise truthfully that the cost of ownership has been reduced. Maintenance cost reduction can be achieved by, for example, the use of a metal repair putty which, in some applications, permits repair and full reinstatement of damaged plant on site without need for dismantling and subsequent welding build-up of the damaged part. Steel car bumpers are increasingly being replaced by plastics, which are displacing many other metal parts to reduce total weight, manufacturing cost and in some cases to offer improved functions.

Tool steel is increasingly being replaced by ceramics such as boron nitride or sialon to ensure longer life and to reduce sharpening and other maintenance costs. Precious metals are, of course, a prime target for the cost reducer; in the telecommunications industry strenuous efforts have always been made to replace electroplated gold for electrical contacts by cheaper materials such as palladium−nickel or palladium−silver alloys, to mention but two. Even in the field of soldering in the electronics industry, increased prices of tin forced modifications to the composition of lead−tin solders by the addition of other alloying elements, such as antimony, to provide similar performance at lower price. A final mundane example is of plastic-based hot-melt adhesives; some are now available in foamed form so that less weight of adhesive material is used for each joint, but with no perceptible reduction in performance.

11.4   Success and failure

The examples offered in the preceding paragraphs are largely of successful substitutions, although some failures have been noted. The reader might wonder how, in his particular area of endeavour, success can be assured. The simple and generalised answer has to be that failures will always occur despite everything that well-meaning engineering and management can do, usually because a vital factor has been overlooked or the influence of some seemingly insignificant parameter has been wrongly estimated. A way around such mental blocks is to arrange that a number of different people consider the problem at different times, and that they discuss it at one or more design review meetings to ensure that nothing important has been overlooked. Figure

11.2 provides some examples and illustrates factors which influence success or failure in materials substitution. It is incumbent on all concerned with the technical aspects of a substitution exercise to consider the impact which a proposed change will have on all facets of the product, including not only the engineering and manufacture but its marketability and subsequent use. This can be done effectively only by discussion with available experts on materials, ergonomics, industrial design, industrial engineering, production, marketing and maintenance. These experts may or may not be available in-house, but, as with so many other aspects of engineering, it is essential to hold discussions at a very early stage if considering a change, rather than bring people in late to solve problems at a time when they cannot be fully effective.

11.5   Strategic materials and economics

The subject of strategic materials is one which more frequently exercises the minds of Governments than of individual companies in industry, although some examples of the latter will be provided later. Strategic materials can be considered as (usually) metals, which originate in countries outside the main user area and for which supply lines are tenuous. The originating country's political situation may be worryingly volatile.

To name a few strategic metals in alphabetical order; beryllium is mainly available from Brazil, India and the Republic of South Africa, (which also has the largest known reserves of chromium) while the Soviet Union, Turkey and Japan are also major suppliers. Cobalt comes from Zaire and Zambia and

Fig. 11.2. Success and failure in substitution. (Copyright: 1984 Standard Telephones and Cables plc. By permission of the Institution of Production Engineers.)

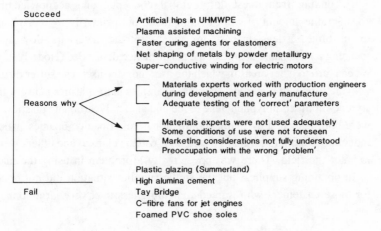

manganese originates in Gabon, Brazil, Australia and the Republic of South Africa. Nickel arrives primarily from Canada, Norway and New Caledonia, while tungsten is supplied by Canada, Bolivia and Korea.

The general problem of availability of strategic materials varies very much with the geographical position of the onlooker. Thus, for example, Gray (1980), Fink (1980) and Jacobsen & Evans (1983) all consider the problems of supply from the American viewpoint, and in that context are not worried that many strategic metals come to them from Canada, through a simple transaction over a land frontier with a friendly nation. Although both these nations are friendly with the UK, there is always the complication in times of stress that everything has to travel by air or by sea before it arrives in Britain. Anderson (1983) reviews the strategic materials supply not only for Britain but for the EEC of elements of particular aerospace interest, including chromium, cobalt, tantalum, tungsten, molybdenum, hafnium, niobium, vanadium, yttrium, titanium and manganese. Although the product designer will not be specifying these as pure materials in his products, he must be aware of the potential for difficulty in selecting special steels which use these elements as alloying agents, quite apart from the economics aspects.

A word on economics might be in order. Most commentators on strategic materials supply consider the need to stockpile, because of the possibility of disruption by overt or covert political action at the supply end, or of the difficulties of transportation in wartime. Although the latter viewpoint may be valid, the fact remains that many of the countries supplying these critical materials are at the same time those with an urgent need for foreign currency; being in the real commercial world they are very interested indeed in maintaining sales to bona fide customers. It seems very unlikely, therefore, that they would wantonly cut off supplies to any western world countries proffering an order, and with money in their hands.

Descending from these lofty global concepts, and acknowledging the considerable amount of work being carried out primarily by metal suppliers to substitute, at least in part, for some strategic metals in alloys, yet still retaining a reasonable degree of the performance demanded (Hodesblatt 1982), we can turn to other areas in which the user industry itself can suffer difficulties with the supply of materials. Familiar examples in the plastics industry include styrene-modified polyphenylene oxide and also polycarbonate plastics, which were both launched by major multinational American companies, who were monopoly suppliers for many years until patents elapsed and others were able to enter the field. There was never really a problem in using the materials or in obtaining supplies, but the single-supplier situation did not bode well for those customers who wanted something less than very large tonnages of simple variants to the materials. A risk in such a situation is that a supplier's key manufacturing plant may stop production for economic, commercial or even accidental reasons, thus limiting supplies for a period. Another difficulty can arise if a product's performance depends upon the use of a particular material which may be one test marketed from a long series of development materials. Such a situation has to be correctly assessed and the implications properly recognised and protected by binding contractual arrangements with the material supplier.

In a more simple, domestic area, poor availability of good quality hardwood for making furniture many years ago stimulated the development of chipboard and blockboard and its consequent veneering as a method of simulating expensive woods. These materials at once brought the benefits of better economics and also offered greater strength than was possible with the original timber. Unfortunately, the owner of furniture made of these boards is left with a reduced sense of the article's intrinsic worth.

Our last example of replacement through poor availability in Northern Europe is that where coal-gas for domestic use was superseded by natural gas from the North Sea and elsewhere. Despite the minimal toxicity of natural gas compared with coal gas, the substitution has not been entirely an unmixed blessing. Apart from the anguish caused by fluctuations in price, there have been technical difficulties because a natural gas-air mixture has a faster flame propagation characteristic than coal gas; this means that the flow rate through a gas appliance is higher than everyone has been used to. Smaller-diameter gas jets have to be used to speed up the gas flow and there follow problems of control and the need for modification of every gas-using appliance in the country.

Companies may suffer from vulnerability of supply of key materials to their industry. For example, in the electronic industry of the UK, most of the metal alkyl compounds used for deposition of aluminium and other metals as surface films for semiconductor use are imported from other countries, as are ceramics such as barium and strontium titanates, cobalt oxide and others, incorporated as materials for ceramic capacitor manufacture. Metals used in the UK electronics industry including tantalum, tin, lead and the platinum metals are all imported, as are varieties of plastic film based on polycarbonate and polyimide. Optical fibre technology depends upon using imported silica, while the form of pure quartz needed for quartz crystal filters also comes from abroad, although quartz crystal oscillators can now be made using indigenous synthetic material. Semiconductor grade silicon is not available in the UK in useable slices, except for experimental purposes, which is also the case with some important process gases such as helium.

11.6   Effects on production processes

A word of warning is justified on the impact that a material substitution in a product may have on the ultimate production process. Although the new material may have been chosen to yield a revised product which appears to offer consistent performance in the customer's hands, seemingly insignificant changes might have a severe effect upon the production rate, costs, or even machinery which must be invoked by the manufacturer. Figure 11.3 illustrates some of the factors in substitution which might impinge upon production processes, and shows some well-known substitutions where minor or major changes had to be made. The thinking designer will, from his own experience and from the examples offered, be instantly able to create a mental image of the changes needed in the production transition between an original material and a replacement. It does not require much expertise to visualise the differences between a production line making automotive carburettors from pressure die-cast alloy, and one making the equivalent product from injection moulded plastic. The consequence is that the designer must confer with production engineers in his early considerations on substitution, so that he can take adequate notice of factors affecting the operations. These include the cost of changing production tooling and machines and changes in manufacturing cycle times.

That the time is right and receptive for significant materials substitution in products is underlined by Waterman (1984), who discusses the four main categories of materials (materials as sources of energy, materials as effect (sic) chemicals, primary role materials and secondary role materials), and also assesses briefly sixteen case histories of new material and manufacturing process exploitation. He concludes that when demand for a product is expanding and

Fig. 11.3. Impact of substitution on the production process. (Copyright: 1984 Standard Telephones and Cables plc. By permission of the Institution of Production Engineers.)

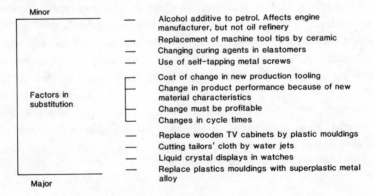

new production facilities are required, the opportunities for applying new materials and new processes are at their greatest; it is very difficult to justify much investment if either product demand or market share is declining.

Finally, the message of this chapter can simply be reiterated. Any material used in a product will offer a combination of properties (the 'envelope') which is exploited in the design. If, for any of many reasons, that material has to be substituted by another, the replacement material will share only some properties with the original one, and even then the coincidence of numerical values may not be exact. Accordingly, the performance of the substitute material in the product will be different, so the substituted product itself may perform in somewhat different ways from the original; for example, it may show a different and unacceptable response to elevated temperatures. The substitution of one material by a second may involve a different manufacturing route, even if only affecting cycle times and temperatures, and this has an impact on production engineering. Furthermore, it may happen that other properties inherent in the replacement material can be exploited in the product to offer new performance characteristics which the originally chosen material could not provide. This then, in turn, should call for a reappraisal of the basic design, which again may lead to product simplification and in its turn have a knock-on effect on the production process. The net result, however, in a well-run operation might yield a revised product with an enhanced performance and a better chance of success in the market place.

12

Material forming processes and design

The product designer could not exercise his profession properly without an extensive knowledge of the processes by which materials are transformed into products. Through this knowledge of processes he can optimise the advantages stemming from them, and avoid at least the greatest of their limitations. The purpose of this chapter, therefore, is to advise the designer of some more recent methods of materials forming, which he may care to study further and subsequently exploit.

A brief reminder of the well-established orthodox forming routes for metals and polymers is given in the next two Tables. Table 12.1 lists some common methods for forming metals, being categorised according to whether the metal being worked upon is molten, heated, or operated upon at essentially a low (room) temperature. Table 12.2 with common forming methods for shaping plastics, lists a few processes which depend upon liquid reactive precursors before a final solid part is obtained, and also mentions methods requiring preformulated material such as moulding granules, stock shapes such as sheet, rod and tube, or simply liquids or solutions.

This chapter considers in turn recent production processes for metals, polymers, composites and ceramics, and concludes with a brief note of some modern techniques with more general application.

12.1   Metals

Extending in considerable detail the information of Table12.1, Phelan and Wyatt (1981) provide a useful summary of specific variants of the main metal forming processes comprising sand casting, die casting, investment casting, centrifugal processes, powder metallurgy and even the deposition of metal layers from fluids. They also discuss forging, extrusion and other processes, with a tabulation of advantages and limitations for each. These processes in recent times have attained a high level of sophistication. For example, CAD−CAM is used for sheet metal working in the USA, by feeding into a computer data on the type of material and its thickness that would preferably be used, and inserting formability data about the materials frictional values relative to the die lubricants and moulds used on the factory floor. The computer then determines whether the part configuration proposed is feasible for the material selected. The calculation can be iterated to suggest alternative design modifications or material substitutions until a part capable of shaping is designed. Brief information on this development, by General Motors in the USA, is provided by Wang and Ayres (1980).

Net-shape finishing is, of course, the ultimate goal in manufacturing to eliminate waste of material and overuse of processing energy. Super-plastic forming of titanium sheet is a recent innovation in the aircraft industry. British Aerospace plans to have a fully fledged production process available by 1987, in which sheets of titanium are clamped in a mould and heated under pressure to over 900 °C. Argon gas blown between the sheets separates and stretches

Table 12.1. *Common methods of forming metals*

With molten metal (melted prior to working)
        Sand casting
        Die casting
        Rheocasting
        Centrifugal casting
        Continuous casting
        Shell casting

With hot metal (heated prior to working)
        Forging
        Rolling
        Swaging
        Drawing
        Extrusion
        Powder metallurgy

With cold metal (becomes heated as a result of working)
        Forging
        Impact extrusion
        Bending
        Piercing
        Spinning
        Hobbing
        Machining
        Rolling
        Drawing (e.g. wire)

them, and forces them into the mould shape while the clamped edges join by diffusion at the same time. In the USA the McDonnell Douglas Corporation prefers to heat this metal with an infra-red quartz lamp, so that the tooling and perhaps the press are not themselves heated. This technique considerably reduces cycle time and saves enormous amounts of energy, Anon. (1984c).

The conventional powder metallurgy route to net-shape parts involving hot or cold pressing in a mould, or pressing isostatically with subsequent sintering, is now supplemented by newer methods. In one, a discharge of 10 kV of electrical energy through a column of metal powder encased in a container effects consolidation in about 50 microseconds, as discussed by Al-Hassani and Sharkery (1980). Theoretically, compacting occurs by interparticle welding, as shown by using low voltage discharges when particles weld into fibres or strands along the current path. As the voltage increases, the number of strands increases and transverse welding takes place. It is claimed that this process makes successful compacts with iron, low alloy steel, stainless steel, high speed steel, nickel and nickel alloys. However, other metals such as cobalt and molybdenum form very brittle and fragile products. Another process patented in Europe involves production of high density compacts based on iron powder by mixing the grains or by coating them with a low melting metal additive based on tin or lead alloys. Heat and pressure converts the low-melting alloy to a liquid phase, and on cooling the compacted product has a density of about 80% of theoretical. If the material is briquetted and sintered, a minimum theoretical density of 97% is achieved.

Another recent method of net-shaping metals involves vapour forming, in which metal is deposited from a compound vapour onto a heated surface. Most familiar is the nickel carbonyl process which has been used for many years to make pure nickel balls for electroplating anodes and for other purposes. However, deposition can be achieved on a heated mandrel to produce a shell-like form of the metal which can subsequently be stripped off, but deposition rates are only of the order of 0.25 mm h^{-1}. Jenkin (1980) provides some details of the process.

Explosive forming of metals on a commercial scale in the automotive industry was achieved in 1980 in Germany, with a saving of 50% of the investment cost which would have been necessary for hot forming of sheet metal, Anon. (1981c). Explosive compression can also be used for manufacturing shapes from amorphous metals. These are non-crystalline or very finely microcrystalline metals, most often available in ribbon form produced by high-speed splat quenching of a melt. However, the powdered material can be compacted by explosive compression in a gas gun to produce billets, tubes, plates and sheets for subsequent processing. The Battelle Columbus Laboratories in Ohio have used a gas-driven, lightweight plastic projectile which punches a powder-filled mould or die at $700-1400$ N mm^{-2}. This produces near-net shapes weighing several hundred kilograms from metal powders, Constance (1981).

Casting and forging are relatively well-understood processes of considerable antiquity. However, some modern variants are now in use. G.K.N. Sankey Ltd in the UK operate a 'Squeeze form' process from which finished aluminium alloy components are produced in a single process by metering the molten metal to match the required volume of the component, and then pouring this into a preheated die cavity in a hydraulic press. On closing, most solidification occurs under pressure and the molten metal thus conforms to the final designed shape. The production rate for this process, which combines gravity die-casting with closed die forging, can be two or three times higher than pure gravity die-casting, and holds the possibility that reinforcing silicon carbide whiskers can be included in the metal to increase strength, see Das and Chatterjee (1981). The 'Injectalloy' process by Remington Arms Company Inc. in the USA, is comparable; finely powdered metal mixed with a plastic binder is injection moulded in conventional equipment. After binder removal by solvent extraction or by heating, the parts are sintered at high temperature to a net shape with $94-98$ % of theoretical density.

We can now review some modern shaping processes. Conventional machining with tools or abrasives, either alone or assisted by plasma heating, is well known. It was perhaps not fully realised until the time of Wilms and Aghan (1981), that the workpiece material may not be isotropic, that is, not homogeneous in every direction. In practice, conventional structural steels for example often exhibit microstructural inhomogeneity as a result of the way in which inclusions and the pearlite—ferrite banding are elongated in the working direction of the steel as it has emerged from the rolling mill. Wilms and Aghan made the first systematic study of the effect of anisotropy on machining and mechanical properties of steel plates. The machining characteristics were shown to be variable and demanded close control to achieve optimum surface finish.

Ultra-high-speed machining of metals has been used on a limited basis in the aerospace industry since the Second World War. Current practice is to effect drilling or milling at spindle speeds of 20 000 rpm with a feed rate of up to 2.5 m min^{-1}. This process can, of course, considerably reduce production costs provided the production volume is high, but care is needed to attain a good surface finish. Brief mention may be made to chipless finishing, more traditionally referred to as roller burnishing. This process cold works the metal surface to produce a very uniform, dense, low centre-line-average finish. Metal is not removed but merely pushed around, so that no chips are produced and the working conditions remain clean. Roller burnishing applies high forces which slightly exceed the yield strength of the workpiece, causing plastic deformation and metal flow.

In another context, plasma-assisted metal cutting has already been mentioned. Very hard steels can be softened locally on the surface to a very small depth by application of a plasma immediately before a conventional metal cutting tool is applied. The tool is easily able to machine away the softened metal, yet the main mass, because of its high heat capacity, does not become sufficiently warm to reduce its degree of hardness. Hot machining allows greater economies with cast or forged parts, especially on those surfaces with sand or slag inclusions. Plasma can also be applied in a carburising process originating from the USA. Whereas in conventional carburising a steel component is heated in a mixed gas atmosphere for eight hours or more to achieve the necessary penetration of carbon, plasma carburising uses a low-pressure hydrocarbon atmosphere within an electric furnace, and the piece part is used as a cathode. Development of a plasma by applying a voltage requires only about 30 minutes of treatment and diffusion to produce a 1 mm deep case on the surface.

Metal surfaces can also be treated with a pulsed laser. Here, the microstructure of the surface is modified by its rapid solidification after melting. McCafferty, Shafrin and McKay (1981) report use of microsecond laser pulses in a vacuum applied to $Al-1.2$ Mn alloy to cause dramatic changes in the surface microstructure.

Finally, in this section devoted to high energy input processes, there is a high-speed deburring process applicable to oxidisable metal parts. Burrs resulting from the machining of small parts are oxidised explosively in a matter of milliseconds when the parts are placed in a pressure vessel, sealed, and charged with a mixture of oxygen and a fuel gas. At the appropriate pressure a spark plug ignites the mixture, and the resulting fire burns for about two milliseconds during which time enough heat is produced to oxidise the protuberances on the parts but not the parts themselves. This process has been taken to the machine development stage.

Still on the topic of surface modification the past few years have seen considerable development of an ion implantation process as a means of imparting surface hardness to metals, with consequently modified frictional properties and, very often, reduced wear characteristics compared with the untreated material. The process is a proprietary one, often applied to rolling or sliding parts and also to cutting tools.

Two recent developments in bonding of metals are worthy of mention. Diffusion bonding has already been alluded to in the context of super-plastic forming of titanium sheet by the aircraft industry. German work has demonstrated that stainless steel can be joined to aluminium by a similar process; properly cleaned surfaces of the two metals are pressed into intimate contact under conditions where a metallurgical bond is formed on an atomic scale, with movement of atoms of one metal across the surface to form interstitial inclusions in the other metal. By this technique, deformation of the material is not essential for joint formation. It is even possible to join alumina to molybdenum, provided that a thin titanium interleaving sheet is used. Failure of the bonds by destructive testing arises as a result of tearing of the bulk metal. Recent information on the process is available, Anon. (1983*d*).

The other noteworthy process involves the use of a ceramic- coated backing ring which can be inserted into pipes to facilitate the making of sound butt-welds using common arc welding techniques. In conventional processes, the most difficult stage is the initial root run in the weld, as subsequent resistance to fatigue in service depends on the condition of this first join, the production of which is highly skill-dependent. Backing rings, at the point of joining, are able to support the molten weld metal, make a good joint and thus provide a fatigue performance equivalent to normal unwelded steel. The rings can subsequently be removed.

12.2   Polymers

Table 12.2 illustrates some of the better-known general processes for converting polymers into finished articles. Turner (1979) provides more complete details of these methods and of many others, listing advantages and disadvantages for each, sufficient to enable a preferred processing method to be chosen. Additionally, for making simple plastic shapes, forging or stamping the materials from premade hot billets or from sheet can often be effected with economic advantage over conventional moulding methods. Grooves and symmetrical radial profiles can often be applied to a simple moulding by centreless grinding; this process can be applied not only to plastics but to some rubbers.

The economics of plastic processing should always be in the designer's mind. West (1984) offers some simple methods of calculating the final cost of a component, depending upon the various cost elements which are necessary to make it. These comprise capital and operational costs. Capital cost obviously involves tooling and plant, and operational costs involve tool setting, product operation and maintenance as well as power used in its lifetime. When a choice is demanded between a number of alternative production methods, this form of costing should be undertaken as a guide to the way forward. However, the capital cost of an injection mould might be quite difficult to establish, although guidance is provided by Walshe and Lowe (1984), who advocate use of a computer to produce quick and accurate mould cost estimates for machining, fitting and polishing stages.

As with metals, lasers can be used for some aspects of plastics processing, since at far ultra-violet wavelengths a high intensity laser causes break-down of the chemical bonding of many polymers which vapourise, taking the heat with the vapour. This is a useful method of removing small amounts of plastic by 'machining', although it was first developed in the USA for etching photomask materials used in semiconductor device manufacture.

The topic of plastics assembly, that is, the production of a component by joining a number of separate piece parts, could fill a volume in its own right. Several plastics material suppliers issue handbooks on assembly methods, which can include various forms of hot or solvent welding, riveting, press and snap fits. Hartley (1980) offers a fairly comprehensive description of these techniques and their variants. For example in welding, one can use spin welding, vibration welding, or hot gases, plates, wires, or electric fields to warm the material to its softening point so that joining can be effected.

Table 12.2. *Common methods for shaping plastics*

From reactive precursors
      Reaction-injection moulding
      Reinforced polyester (for GRP)
      Monomer polymerisation
      Transfer moulding of thermosets

From preformulated material
      Melt casting
      Injection moulding
      Extrusion
      Compression moulding
      Transfer moulding
      Rotational moulding
      Powder spraying

From stock shapes
      Thermoforming of sheet
      Machining
      Calendering of sheet
      Heat-bending

From solution or suspension
      Spin-casting
      Slush moulding
      Knife coating

12.3   Composites

This discussion on the forming processes for handling composite materials will be confined to those based on polymers. Of these, by far the most familiar are the glass-fibre reinforced polyesters based upon a reactive monomer catalysed to convert to a rigid solid. The process of incorporating glass fibres is familiarly known as 'laying-up', and originally was carried out by laying glass fibre on a mould and impregnating it with syrupy resin by pouring or spraying. Subsequent catalysis and heat curing then produced an article which could be of considerable size, such as a boat hull. Variants of this technique have involved spraying on the resin while simultaneously feeding chopped glass fibre into the gun nozzle; others require use of preprepared glass-cloth tape as a reinforcing medium, or the winding of filaments of reinforcing fibre around a mandrel with simultaneous addition and curing of the liquid resin.

Other processes with similar types of material involve operation on dough moulding or sheet moulding compounds; these are essentially preprepared doughs available in bulk or as preformed sheets, made up of a partly cured resin with glass or other fibre reinforcement already incorporated. These materials are moderately dry to the touch and so are much easier to handle than the liquid types. They are fabricated by compression moulding in a hot press or by extrusion which, when traction is applied to the product and a simultaneous cure process applied, is known as 'pultrusion'. With all these materials, a heating process has hitherto been necessary to effect catalytic cross-linking of the polymer to provide a rigid structural material. However, recently, new chemicals have become available which enable these products to be manipulated up to within a few seconds of their final shaping. The basic materials are supplied as sheet from a continuous production line, and will contain up to 60% of glass fibres, being supplied on a roll which can be stored in the dark for long periods. However, on exposure to ultra-violet light, even if only that from the sun, curing to a state of rigidity occurs in only a few minutes, and the material is then hard enough to be handled and further cured by more powerful and controlled ultra-violet light exposure. A restriction, of course, on this material is that it must be sufficiently transparent to ultra-violet light, within the required thickness of section, that underlying levels of material are also cured.

Machining of fibre-reinforced composites has always posed difficulties, because of the problems of chip formation and the cutting forces which tend to expose the ends of fibres and, to some extent, drag them from the matrix, thus causing roughness and interfacial weakness in the material. A little has been published, however, on carbon-fibre-reinforced plastics (CFRP) and the machining which is needed for making holes and trimming edges. It has been found, particularly with CFRP, that life is made difficult by delamination of composites and the short tool lifetime. Koplev, Lystrup and Vorm (1983) have recently contributed to our understanding of the cutting process for CFRP by their machineability studies. Bearing in mind that, previously, even carbide tipped tools are often worn out when less than one metre-length of hole had been drilled in some of these abrasive materials, it was important to investigate optima for the rake angle, relief angle and cutting depth achievable. If the composite was machined parallel to the fibres, the surface was smooth, and machining cracks reached only one or two fibre diameters into the composite. However, these cracks would be as long as 0.3 mm if machining was in a direction normal to fibre orientation. Such considerations, of course, considerably reduce the usefulness of machining as a process for shaping any other than geometrically favourable products based upon such a composite.

Carbon-reinforced thermoplastic composites enjoy favourable joining characteristics because they will conduct sufficient electric current to be bondable by resistance welding. Work in the USA has shown that, in the case of carbon−polysulphone rods, a heating time of two seconds and passage of up to 90 joules of energy was sufficient to effect a useful join. Of course, in these composites, conventional bonding by drilling a hole and using a bolt is to be deprecated owing to the generation of minute radial cracks around the drilled hole; these can very quickly lead to fatigue failure in service.

12.4   Ceramics

Chapters 3, 5 and 8 in the sections dealing with ceramic materials should have made the reader familiar with the more common production processes for these materials, so further explanation is hardly necessary. Conventional sintering technology developments tend to be aimed at reducing process time and at increasing final density of the compacts while maintaining predictable shrinkage in defined orientations. A newer alternative process involves explosive compaction. Although the first applications of explosives were reported as early as 1872, explosives used to compact powders are of more recent interest. Ceramic powders have been explosively compacted to high density at pressures up to 7 GPa, as shown by Koenig and Yust (1981). Microstructural examination of the compacted and subsequently annealed ceramics showed that even when densities as high as 90% of theoretical were achieved, microcracking was prevalent and dislocations and massive damage faults were evident, although interparticle bonding and even some melting was seen. In the cases of aluminium nitride, silicon nitride, boron and alumina, the opinion was formed that although cracking is present, microhardness measurements indicate some useful strength and structural integrity. This type of work is at an early stage, and industrial potential still seems far away.

Another recent treatment process for ceramics, paralleled by development in metals, is ion implantation. In the case of, for example, tungsten carbide cutting tools, life in service is usefully increased, probably because the implanted atoms strengthen the cobalt matrix which acts as a binder for the ceramic particles. This type of process has been widely published at least in the UK, and is available as a commercial service.

12.5   Newer production possibilities

We conclude this chapter by describing some recent production developments involving materials, which might be exploitable over a wide range of products by designers with an eye for fresh opportunities.

Contactless manipulation and processing of small articles is, in principle, possible by using an acoustic wave levitation system developed by the Jet Propulsion Laboratory in the USA. In this system, a long hollow trough is used as an acoustic focusing radiator, being fed from ceramic piezoelectric transducers closely coupled to it to generate acoustic energy in the form of a line along the trough's axis. Enough energy is present here to levitate liquid drops, millimetre size microballoons and other small articles. This implies that such items can be subjected to chemical or other vapour deposition processes without need for contact with the otherwise supporting surface, where such contact might upset geometry or purity. This equipment has been used to produce small metal and glass components. Ultrasonics may also be used in the liquid phase, for example, to stimulate deposition of metal in an electroplating process. Drake (1980) reports the electrodeposition of gold in which high-frequency ultrasonic radiation is used to enhance metal deposition on preselected areas of substrate to be used for electrical contacts. The deposition rate in the irradiated area is about three times that which occurs in the rest of the bath so that use of precious metal can be minimised as it is deposited preferentially in what is technically the most important area. Due to the complexity of the equipment, it is unlikely that such a process could be used to significantly enhance deposition in most commercial large-scale plating operations using base metals.

Lastly, we can mention the use of high-pressure water jets for cutting. A fine jet of water fed through a small nozzle at 70 kN/m^2 can have a considerable cutting effect on soft materials such as cloth, foamed plastics or even bread. It has recently been reported, Anon. (1984*b*), that putting zircon sand in the water provides an effective abrasive jet which can cut plastics or even glass up to thicknesses of 40 mm. The edge quality provided is similar to that obtained from diamond grinding, and the process avoids surface damage. Cutting has to be started from an edge or a previously machined hole, but seems not to be shape-limited. The process could also be used on materials like titanium, high-tensile steels and plastic laminates. Straight line cutting speeds of 0.5 m min^{-1} are claimed on 12 mm thick glass.

13

Sources of information on materials

13.1 Introduction

The product designer would be the first to admit that in many areas of knowledge he is sadly lacking. It is the usual experience that those who choose materials for products are usually (and rightly) much more knowledgeable about the operation of the product, and the means by which their associates can make it, than about the finer points of materials science or technology, and especially on the question of where to find materials data.

The successful designer manages to survive very well with his inner resources of knowledge, through his contacts with colleagues and consultants, and with his personal collection of well-tried, tested, and favourite books and data sheets. However, the world moves on rapidly, and its materials technology activity generates a wealth of information. Access to this is increasingly vital to every company in the present climate in which research, development and product lifecycles become ever shorter.

Although several texts, in passing, assist the reader to find information, few of them treat this aspect of a product designer's operations specifically. This chapter seeks to fill that gap by focusing upon the various sources of materials information available at present. Although primarily concerned with the UK some reference is made to overseas sources, in particular via directories which offer information on world-wide data bases.

13.2 Location of documents

The most obvious sources of information for the neophyte are libraries. Any library holding scientific or technological information will make it available as open and unclassified literature. However, many will not permit removal of a book from the building; some indeed require that a book is requested specifically at the reception desk before it is withdrawn from shelves not available personally to the intending reader.

Most libraries hold a selection of information sources ranging from technical dictionaries, encyclopaedias, text books and monographs, to bibliographies, indexes, abstracts, technical and professional journals, technical reports, and sometimes catalogues and manufacturers' brochures. Dumper & Loader (1984) provide a comprehensive list of important British libraries with addresses and telephone numbers. The most comprehensive libraries for scientific and technological purposes in the UK are the Science Reference Library in London and The British Library, Lending Division, in Wetherby, Yorkshire. The City Business Library in London mainly houses commercial, but also some technical, information.

Some libraries deal specifically with metallurgical or engineering subjects, these amongst many others being listed by Lingenfelder (1983). Another useful book by Codlin (1982), refers to materials science, plastics and ceramics, but not in finer detail to the remainder of materials technology. It lists commercial companies, addresses, telephone numbers and subjects. Lastly, Harvey & Pernet (1981) list those organisations in Europe which handle scientific and technical subjects.

Supplementing the major city libraries are those in universities, polytechnics and colleges, about which some detail is provided in Section 13.4, and also the libraries of professional institutions, research and trade associations and Government organisations. In general it is necessary to make a prior appointment before visiting any of them.

13.3   Published information
13.3.1   *Journals*
All technical libraries offer a range of scientific, technical and trade journals dealing with materials, the degree of coverage depending upon local demand and upon the available budget. Journals range from the highly scientific to those more easily read by the average designer; at the other end of the spectrum are trade bulletins providing production data by country or by industry, and also news of industrial successes and manufacturing changes. Table 13.1 gives a subject-grouped list of journals and is offered as a representative selection.

13.3.2   *Books and bibliographies*
Books range from dictionaries and encyclopaedias, useful for background reference, through engineering handbooks which present specific methods for making strength, heat-flow or other calculations, to text books covering the whole possible spectrum of reader interest and education. Specialist text books may sometimes prove too complex and detailed for a designer, but a bridge is provided by interpretative books such as *Annual Reviews of Progress* in selected areas of science and technology.

Table 13.1. *Some useful materials-oriented journals*

Ceramics and Glass	*American Ceramic Society Journal* *British Ceramic Abstracts* *British Ceramic Society Journal* *Cement and Concrete Aggregates (ASTM)* *Glass* *Glass Industry* *Glass Technology* *Journal of American Ceramic Society* *Magazine of Concrete Research* *Physics and Chemistry of Glasses*
Composites	*Composites Science*
Corrosion	*Anti-corrosion Methods and Materials* *Corrosion* *Corrosion Science*
Design	*Design Engineering (London)* *Design Engineering (USA)* *Engineering Materials and Design* *Eureka* *Materials and Design* *Materials Engineering*
Fabrication	*Journal of Applied Metalworking* *Powder Metallurgy* *Processing* *Welding Journal*
Metals	*Brazing and Soldering* *Bulletin of Alloy Phase Diagrams* *Japan Institute of Metals, Transactions* *Journal of Less Common Metals* *Journal of Metals* *Metal Bulletin* *Metal Progress* *Metal Science* *Metallurgia* *Metallurgical Transactions* *Metallurgist and Materials Technologist* *Metals Technology* *Metals and Materials* *Powder Metallurgy International* *Tin and its Uses*
Miscellaneous	*Chemistry and Industry* *Fire and Materials (ASTM)* *Journal of Electronic Materials* *Materials in Engineering* *Materials Letters*

Table 13.1. (*cont.*)

	Materials Performance
	Materials Research Bulletin
	Wear
Plastics	*British Journal of Polymer Science*
	British Plastics and Rubber
	European Plastics News
	European Rubber Journal
	IEEE Transactions on Electrical Insulation
	Insulation/Circuits
	Journal of Applied Polymer Science
	Journal of Polymer Science
	Modern Plastics International
	Plastics Design and Processing
	Plastics Engineering
	Plastics and Rubber International
	Polymer
Science	*Chemistry in Britain*
	Current Awareness in Particle Technology
	Journal of Materials Science
	Journal of Non-crystalline Solids
	Materials Science and Engineering
	Progress in Materials Science
Surface Finishes	*Metal Finishing*
	Plating and Surface Finishing
	Product Finishing
	Transactions of the Institute of Metal Finishing
Testing	*British Journal of Non-destructive testing*
	Non-destructive Testing International

Note: Full details of all these journals, and of thousands more, are held in 'ULRICH' (1984) and are available online.

Wimberly (1985) provides a useful summary of key sources of technical information in the USA, slanted to the needs of the engineer.

Further downmarket are trade directories, which often embody technical sections, considerable advertising matter, and classified lists of suppliers of materials or of services.

Some useful abstract titles covering materials topics are:

Applied Science and Technology Index.

British Ceramics Abstracts.

Chemical Abstracts.

Corrosion Control Abstracts.

Current Awareness in Particle Technology
Metals Abstracts.
Engineering Index

Searching published literature is an art rather than a science; not something to be done effectively by the casual student. A good technique, having found the UDC (Universal Decimal Classification) or locally used shelfmark for the subject of interest, is to locate the most available up-to-date book or reference source on that subject and then to check its bibliography to back track to more specific, though earlier, detail. Unless one is interested in historical research, it is seldom necessary or even desirable to go further back in time than ten or fifteen years when reviewing materials subjects, if the intention is to locate information about currently available materials.

Details of suppliers of materials are held in specialised suppliers' data books. These have a common format of indexes to products and services, with separate details of suppliers' addresses. For example, information about the trade aspects of surface finishing metals is available in *Metal Finishing*, (1984).

13.3.3   *Directories*

Directories are books, not in themselves educational or directly readable, which indicate sources of required information. A good introduction to this literature comes from Anderson (1985) who provides information about directories covering subjects such as plastics, metals and ceramics. Another regularly updated publication, dealing with current research projects in British academic institutions, derives from Pilling (1984).

Associations are sometimes a useful source of advice on materials; Henderson & Henderson (1980) and Millard (1971) have produced directories listing them. The Science Reference Library in London, holds a very large collection of trade directories for the UK, the Commonwealth, Europe, North America, Africa and Asia – Australasia.

When searching for consultants, the designer can be guided by Liddiard (1980). Equally compendious is the annual *Directory* published by the British Consultants' Bureau of London, mainly to promote the work of British consultants overseas.

The Department of Trade and Industry publishes a booklet *Technical Services for Industry*, providing condensed information on British Government and academic laboratories, details of their current research programmes and names of contacts in them. Anon. (1981*d*), provides details of members of research establishments, but offers no keyword listing of their special interests; however, the book can be used to check the credentials of candidate consultants found by other means.

The needs of the materials purchaser are covered by many directories of suppliers, of which the *European Plastics Buyers' Guide*, published annually by IPC Industrial Press Ltd, London, is an example.

The reader interested in sources of scientific documentation beyond the scope of this book is recommended to consult *British Scientific Documentation Sources* published by the British Council.

Sometimes, a group of people organises itself as a purveyor of data. For example, the Metals Information Group comprises the information departments of seventeen leading UK organisations concerned with all aspects of metals production, processing and use. The group publishes a directory of the main metal organisations in the UK, with further information on their library services, their expertise and contacts for special services. It is available from the British Steel Corporation in Sheffield.

Lastly, mention must be made of directories of materials research centres operating in the UK, Europe or the world. UK research activity is well covered by Williams (1980). The European scene is reflected in a companion volume, Williams (1982). Mitchell & Lynes (1983) offer a world directory of organisations and programmes in materials science. Obviously, these books reflect only those positive returns made in response to a generally broadcast questionnaire, but nevertheless together they are a very useful starting point for further information, particularly as telephone numbers, names and addresses are provided.

13.3.4   *Conference proceedings*

Conference proceedings constitute a useful, if somewhat specialised and perhaps ephemeral, source of written information reflecting selected facets of the current state of science and technology. Inevitably they mirror individuals' narrow viewpoints, so proceedings are useful primarily as a supplement to other sources of information, rather than in themselves constituting working documents for the materials selector and product designer. Conference proceedings are usually mentioned in abstracting journals, and full copies are held in many libraries. The intending user should understand the local library procedure for identifying them, as title, locale and dates do not neatly fit into the normal journal reference systems, and proceedings may be held in a separate section of shelves. It is frequently important when tracing proceedings to know the name of the town in which the conference was held; this has at least the side effect of teaching the researcher some American geography, as the USA is by far the most favoured venue for materials conferences.

13.3.5   *Films*

Films and video-cassettes originate from many sources; suppliers of materials extol their wares, pressure groups put forward points of view, government departments justify their activities, academics seek sponsors and commercial enterprises seek customers.

There appears to be no central point which advertises film availability. Many libraries hold catalogues, and some groups and associations, such as the British Plastics Federation, the Chipboard Promotion Association, the Building Research Establishment and the Welding Institute make and distribute films about their work.

13.3.6   *Patents*

Most civilised countries operate a patent system, effectively offering the proprietor monopoly rights over an invention for a specified time, in return for which he has to disclose full details. However, because a patent is intended to be primarily a legally enforceable document, emphasis focuses on its legal standing rather than on its value as a technological data source.

It is unlikely that those unskilled in the art (to modify the conventional phrase) would be able to take a patent specification and, from nothing else, reconstitute the invention, although there is more possibility of this happening with US patents, which are required to be much more informative than those from other countries. However, the technological information contained in patents, particularly relating to materials and their processing, does provide a good account of what is currently known, even though it will never be so up-to-date as the open scientific literature because of delays inherent in official patenting procedures. 80 % of the information in patents never appears in any other published form, so their use as a technological resource should not be underestimated.

The Science Reference Library, London, houses patents from the UK and from many foreign countries. The patent specifications themselves need not necessarily be found and studied as a weekly journal of abstracts can be consulted; the *US Gazette* is the American equivalent of this abstract publication. For Japanese language patents, Derwent Publishing Company provides a computer data base which can be accessed, so that at least abstracts may be read in English. This Company handles patents from other countries, as well.

13.3.7   *Computer data bases*

The computer, with its enormous memory capability, is obviously useful when seeking information on materials. Searching published literature has always been recommended as an essential preliminary to undertaking a project, be it an advanced scientific work or the development of an engineered device. Unfortunately, because of the explosion in the quantity of information over the past 20 years, searching through printed literature is now often neglected, as the prospect of looking through sufficient indexes seems too daunting.

These days the computer is available, and when terminals are provided designers or project engineers can now enlist the services of a skilled operator or intermediary and by working together can carry out a computer search of selected parts of Data Bases holding millions of items.

The main search operations involve selecting keywords to express the requirements of the questioner and then connecting a local terminal to the main frame computer; after selecting the file to be searched, keywords can be entered into the system. The responses, in the form of appropriate references, are printed out for further visual selection by the questioner. Keywords arise from a number of sources; they can come from the title or from an abstract, but the term also embraces descriptors (usually controlled indexing terms), identifiers (uncontrolled indexing terms), authors, the corporate source of authors, classification codes, registry numbers and document types.

Those public libraries able to offer such a search service are listed by Aslib (1981). Educational establishments can sometimes offer the services of an intermediary; the library at Hatfield Polytechnic, Hertfordshire, is a case in point: in a private company, it is hoped that a computer enthusiast can be enlisted for the task.

Data bases of most use to the enquirer interested in materials, and available from a number of on-line companies, are shown in Table 13.2.

Computers can also be used for selecting materials. This is generally effected by comparing materials property data held on file against an enquirer-generated performance specification; the computer simultaneously identifies those materials which meet the requirements. Work by Imperial Chemical Industries on a very wide range of plastics in the field of computer selection has been reported by Pye (1983*b*), who also offers information on the computer-aided adhesive selector (CAAS) developed by the Permabond Adhesives Company. CAAS uses a simple question and answer procedure to guide the user through 20 options which lead him towards the most suitable type of adhesive.

For plastics selection, RAPRA Technology Ltd, Shawbury, Shrewsbury, Shropshire, SY44 NI, recently announced their plastics materials selection system PLASCAMS 220. This menu-driven system holds data on all currently available generic plastics and simplifies the user's performance specification to a short list of candidate materials.

Table 13.2. *Some important materials-oriented data bases*

Data base origin or name	Information available
1. *BLAISE*	Blaise Marketing, 7 Rathbone Street, London W1P 2AL
UK MARC	Information on books and new serials catalogued by the British Library for the current period. This includes copyright deposited and also bought material
UKRMARC	As above, for earlier data
LCMARC	Library of Congress machine-readable cataloguing data base
Conference Proceedings Index	Marketing Office, British Library, Bibliographic Services Division, 2 Sheraton Street, London W1V 4BH
2. *ESA-IRS*	B.A. Kingsmill, Dialtech Service, Technology Reports Centre, Orpington, Kent BR5 3RF
Aluminium	All aspects of aluminium metal from American Society for Metals
Aqualine	Data from the Water Research Centre
Chemabs	Chemistry and chemical engineering
Compendex	Engineering data
METADEX	Metals data on materials, properties, processes and products from American Society for Metals
BNF Metals	Data on non-ferrous metals from UK
NASA	Reports from US National Aeronautical and Space Administration. Data restricted to USA and NASA/ESA contractors
OCEANIC	Materials for marine use
PASCAL CNRS.	Wide area of science including metals

Table 13.2. (*cont.*)

Data base origin or name	Information available
3. *INFOLINE*	Pergamon Infoline Ltd, 12 Vandy Street, London EC2 2DE
RAPRA	Worldwide coatings abstracts, ceramics abstracts, Patsearch, Impadoc, Compendex
Chemical Engineering Abstracts	Royal Society of Chemistry, The University, Nottingham. NG7 2RD. Materials for chemical engineering
WPI	Index of patents from all major industrial countries, from Derwent Publications
ZLC	Abstracts of references to zinc, lead and cadmium Zinc–Lead Development Association and Cadmium Association
4. *LOCKHEED*	PO Box 8, Abingdon, Oxford, OX13 6EG.
BHRA Fluid Engineering	Data on statics and dynamics, laminar and turbulent flow, from British Hydromechanics Research Association
CA Search	Chemical abstracts, with general subject headings and CA registry numbers
CLAIMS	US Patents
COMPENDEX	Engineering and technological literature
Comprehensive Dissertation Abstracts	Subject, title and author guide to US dissertations
Conference Papers Index	Details of conference papers, proceedings and publications, from Data Courier Inc
Foreign Traders Index	Manufacturers, retailers, wholesalers etc. in 130 countries. Data from US Dept. of Commerce

GPO Monthly Catalogue	Details of US Government reports and proceedings from US Government Printing Office
METADEX	American Society of Metals data on metals
Non-ferrous Metals Abstracts	Equivalent to BNF Metals on ESA-IRS
NTIS	Reports of US Government – sponsored research
SCISEARCH	Citation searching possible on the science and technology files of the Institute for Scientific Information
Standards and specifications	Details of USA-originated documents
WELDASEARCH	Joining of plastics and metals. Metal spraying and cutting. Data from the Welding Institute
World Aluminum Abstracts	All aspects of aluminium as metal
WPI	Index of patents from all major industrial countries

Of more general application is MATUS (Materials User Service) offered by Engineering Information Company Ltd, 30 Hampton House, 15–17 Ingate Place, London SW8 3NS. This data base incorporates suppliers' information about many different types of materials including metals, alloys, plastics, ceramics and refractories, having been taken from published technical catalogues, again for specification-matching purposes. Furthermore, this system can be searched by trade name, manufacturer, composition, application of material or by specific properties.

13.4   Current research

Section 13.2 examined the physical location of information on materials in libraries, and Section 13.3 discussed the different forms in which this information is held and presented. The present section deals with those organisations generating basic materials information from their own research and development.

13.4.1   *Universities*

In Section 13.3.3 on directories, mention was made of books which describe materials and industrial research in the UK. Of these, the most important for our present purposes is by Pilling (1984), detailing current work in most of these establishments. Although almost all the Universities in the

Table 13.3. *Materials-oriented UK universities*

Birmingham University
Bradford University
Brunel University
Cambridge University
Leeds University
Liverpool University
London University (Imperial College)
London University (Queen Mary College)
Loughborough University of Technology
Newcastle-on-Tyne University
Nottingham University
Open University
Oxford University
Sheffield University
Strathclyde University
Surrey University
University of Aston
University of Bath
University of Manchester, Institute of Science and Technology
University of Wales, University College, Cardiff
University of Wales University College of Swansea

UK have departments of chemistry, physics, engineering and cognate subjects, only those listed in Table 13.3 have faculties specialising in materials science or technology.

Many universities operate an Industrial Liaison Office primarily to organise client-university contracts intended to earn money, or at least to make some of the university projects financially self-supporting. For this reason, Industrial Liaison Offices are not geared to answering technical questions from casual enquirers, but they will always point the way on request to university staff members.

13.4.2 *Polytechnics*

As with university faculties, some polytechnics have significant research involvement with materials, and some of them also operate an Industrial Liaison Service. They are listed in Table 13.4.

13.4.3 *Colleges*

Compared with the larger academic institutions, colleges are less prominent in the fields of materials research and technology. However, there are some noteworthy organisations listed in Table 13.5.

Table 13.4. *Some materials-oriented polytechnics*

City of London Polytechnic, Sir John Cass School of
Science & Technology
Coventry (Lanchester) Polytechnic
Huddersfield Polytechnic
Manchester Polytechnic
Middlesex Polytechnic
Newcastle-on-Tyne Polytechnic
North Staffordshire Polytechnic
Polytechnic of North London
The Polytechnic of the South Bank
Sheffield City Polytechnic
Teesside Polytechnic
Thames Polytechnic

Table 13.5. *Materials-oriented colleges*

Cranfield Institute of Technology
National Institute of Agricultural Engineering
Royal Military College of Science

Note: These colleges have interests in metals, plastics, composites, ceramics, welding and wear properties of surfaces.

13.4.4   *Research associations*

For more than a century organisations with common technical and commercial interests have clubbed together to set up Research Associations. These exist primarily for technical support of their corporate subscribers, but several carry out industrial research on an external contract basis. All have greater or lesser library facilities which usually can be used by the public by prior arrangement with the librarian. The Associations in Table 13.6 are concerned with materials, names being indicative of the main activities.

These Associations usually produce reports for the benefit of their members or for paying clients, but most do not produce open documentation as a commercial venture. However, the Paint Research Association publishes a list of its regular and special publications which are free of charge to members or sold on a commercial basis to non-members. Other bodies such as RAPRA Technology Ltd not only produce client-confidential reports, but also operate a large business publishing text books, monographs, reports, conference proceedings and data sheets. Of these, particularly interesting to product designers are books by Thorn (1980) and Cooper (1982) in the field of plastics.

The BNF Metals Technology Centre offers commercial publications in the form of guides based on its own research, and also reports on specialist subjects. It publishes conference proceedings, miscellaneous documents and papers on behalf of the British Association for Brazing and Soldering and for the International Wrought Copper Council.

Table 13.6. *Materials-oriented research associations*

BHRA Fluid Engineering
BNF Metals Technology Centre
BABS (British Association for Brazing and Soldering)
BCIRA (British Cast Iron Research Association)
British Ceramic Research Association
British Glass Industry Research Association
Cement and Concrete Association
ERA Technology Ltd
Malasian Rubber Producers' Research Association
The Motor Industry Research Association
Paint Research Association
Production Engineering Research Association
RAPRA Technology Ltd
SIRA Institute Ltd
Spring Research and Manufacturers' Association
The Steel Castings Research and Trade Association
Timber Research and Development Association

13.4.5 *Government organisations and research establishments*

Much of the work from those laboratories listed in Table 13.7 is open to public scrutiny. A question to the librarian will often identify a materials expert within the organisation, and perhaps elicit details of their openly published information.

Names of key staff and brief notes on the activities of these organisations are provided in the regularly updated Department of Trade and Industry Publication, *Technical Services for Industry*.

Some of these organisations, for example the Building Research Establishment, produce extensive bibliographies of reports and publications, obtainable from the librarian.

The United States of America is, not surprisingly, a prolific source of information on research and development; its Federal Government either conducts directly, or pays for, most of the research carried out in the country. A monthly catalogue of all reports available from Federal Agencies is produced by, and can be obtained from, the Government Printing Office, Washington DC 20401, USA. The National Technical Information Service and National Aeronautics and Space Administration Research Reports, and more, are available through the appropriate data bases discussed in Section 13.3.7. In recent years, the US Government has set up a network of Information Centres intended to collect, store, and disseminate world literature for the defined fields in which each operates. The output of the Centres comprises state-of-the-art reports and bibliographies. Of interest to the materials expert is the Metal Matrix

Table 13.7. *Some important materials-oriented government organisations*

Admiralty Marine Technology Establishment
Admiralty Marine Technology Works Admiralty Research
  Establishment
Atomic Energy Research Establishment
Atomic Weapons Research Establishment
Building Research Establishment
Materials Quality Assurance Directorate
National Engineering Laboratory
National Gas Turbine Establishment
National Physical Laboratory
Non-destructive Testing Centre
Propellants, Explosives and Rocket Motor Establishment
Royal Aircraft Establishment
Royal Armament Research Development Establishment
Royal Signals and Radar Establishment
Water Research Centre (Warren Spring Laboratory)
Water Research Centre (Medmenham Laboratory)

Composites Information Analysis Centre at Kaman Tempo, Santa Barbara, California 93102, The Metals and Ceramics Information Centre at Battelle Columbus Laboratories, Columbus, Ohio 43201, The Plastics Technical Evaluation Centre at Picatinny Arsenal, New Jersey, and the Machineability Data Centre, Metcut Research, Cincinnati, Ohio.

13.4.6 *Public corporations*

A comprehensive survey of the operations of UK Public Corporations is offered by Williams (1980). Some of these have a pool of expertise accessible to the public through their research laboratories. Those likely to be of interest to designers of engineered products are shown in Table13.8.

The National Coal Board maintains on a computer details of those materials approved, and of those prohibited, for use below ground in order to control situations which could lead to explosion or fire risks.

13.5   Further information sources

13.5.1 *Professional institutes*

The Institutes listed in Table 13.9 are often qualifying bodies for practitioners, and as such have a detailed infra-structure, usually including a library and often a publications activity, as listed by Dumper & Loader (1984), who also usefully provide a worldwide list of Welding Institutes. Although publications are purchaseable on the open market, their libraries will usually only be accessible to the public on prior application to the librarian. It is not to be expected that any of these Institutes or Institutions will be able to offer laboratory facilities, the appropriate Research Association being the proper place for that activity. All contacts should initially be to the Secretary of the Institution, detailed addresses being available from Williams (1980) or Codlin (1982).

Most of these organisations publish a regular journal including technical papers; they are usually indexed annually and, in some cases, entered into computerised data bases, thus being amenable to keyword searching.

Table 13.8. *Some materials-oriented public corporations*

British Aerospace
British Gas Corporation Research and Development Division
British Steel Corporation Teesside Laboratories
British Telecommunication Research Laboratories
Central Electricity Research Laboratories
National Coal Board Coal Research Establishment

Some of these organisations offer publications for sale. For example, the *Materials Optimizer* offered by Fulmer Research Laboratories Ltd (the Institute of Physics) has already been mentioned. The Institute of Ceramics sells a range of text books and publications from the British Ceramics Society, while the Institute of Corrosion Science and Technology offers monthly publications and various booklet guides and bibliographies. The Institution of Production Engineers offers a book dealing with soldering, brazing, welding and adhesives, and also a tape/slide presentation *The Wealth Creators*, which explains the career and training of a production engineer; it could be useful as background information to the product designer. The Welding Institute provides formal catalogues of publications, films and video cassettes on a wide range of subjects, and during 1984 went into print with a series in *Welding Institute Research Bulletin* under the generic title *How do I find out about . . . ?* Of interest to the designer, are Number 1 *Identification of Steels*, Number 3 *Identification of Non-Ferrous Metals*, Number 4 *General Information on Stainless Steels* and Number 12 *Where to find out*.

Table 13.9. *Institutes and other bodies with materials information*

Association of Technical Institutions
British Institute of Engineering Technology
British Institute of Non-destructive Testing
Cranfield Institute of Technology
Fulmer Research Institute
Institution of British Engineers
Institute of Ceramics
Institution of Corrosion Science and Technology
Institution of Engineering Designers
Institute of Engineers and Technicians
Institution of Gas Engineers
Institute of Metal Finishing
Institution of Metallurgists
Institute of Metals
Institution of Mining Engineers
Institution of Nuclear Engineers
Institute of Physics
Institution of Production Engineers
Institute of Professional Designers
Institute of Sheet Metal Engineering
Institute of Sound and Vibration Research University of Southampton
International Tin Research Institute
Plastics and Rubber Institute
Royal Society of Chemistry
Welding Institute

13.5.2   *Trade associations*

These exist to meet the needs of materials manufacturers and users in particular industries. Organisations listed in Table 13.10 have obvious connections with specialist areas in materials science and technology; in some cases they have libraries which can be visited by prearrangement.

Some of these bodies produce publications. For example, the Aluminium Federation publishes *The Properties of Aluminium and its Alloys*, the British Plastics Federation offers films about plastics, and the Copper Development Association furnishes an excellent series of Technical Notes relating to copper composition, properties and applications in pure and alloyed forms. The Zinc Development Association is well-known for its regularly updated *Galvanizing Guide*.

13.5.3   *Museums*

Some museums are able to offer specialist but, nevertheless, valuable materials information. Those worth contacting through their Directors are shown in Table 13.11.

Table 13.10. *Materials-oriented trade assocations*

The Aluminium Federation
The Association of Metal Sprayers
British Ceramic Society
British Mechanical Engineering Confederation
British Non-Ferrous Metals Federation
British Plastics Federation
British Society of Rheology
Copper Development Association
Engineering Materials and Design Association
Faculty of Royal Designers for Industry
Federation of Engineering Design Companies
Lead Development Association
Oil Companies Materials Association
Society of Glass Technology
Zinc Development Association

Table 13.11. *Some museums with materials interests*

Birmingham Museum of Science and Industry
British Geological Survey
British Museum Research Laboratory
Manchester Museum of Science and Technology

13.5.4 *Other sources*

Three other sources of materials information must be mentioned. The Design Council offers a Designer Selection Service and the Society of Industrial Artists and Designers offers a Design Information Service. Catalogues issued at Exhibitions, especially the larger National ones, usually have advertisements and other mention of suppliers, although such lists are not selected on a rationalised basis for the professional reader.

13.6   Materials suppliers

Suppliers of materials will always produce information on their products, usually in terms of mechanical, electrical or environmental performance. Many operate Technical Service departments in which development work is carried out to customers' requirements. These departments often issue *Design Guides* for their products, thus providing clues to the performance of a whole generic class of materials. Some suppliers issue their own regular house journal on materials; examples are *Innovation* from E.I Dupont de Nemours, and *Plastics in Engineering*, from the Celanese Corporation.

Within some industries the infrastructure of information sources to the customer can be very large. For example, the British Steel Corporation offers a Structural Advisory Service with eight engineers providing nationwide coverage, and a 'hot-line' service to help customers with queries relating to their products.

Other corporations and industries have parallel infrastructures.

Materials suppliers are essentially materials selection consultants operating over a restricted field of vision. Nevertheless, once a selection has been made, and a supplier has been found, full use should be made of any special services offered, including application and prototyping laboratory services where these are available; this enables both the supplier and the designer to obtain hands-on experience of the problem.

Table 13.12. *UK Directories of suppliers and services*

> *Directory of Consulting Engineers*
> *Euro-Pages (Index of Products and Services)*
> *Kelly's Manufacturers' and Merchants' Directory*
> *Kemps Directory*
> *Kompass Directory*
> *Sell's Directory*
> *Yellow Pages (British Telecom)*

One might wonder where the names of materials suppliers can be found. The UK is particularly well served with directories, which are generally sub-divided into sections dealing with products and their suppliers and with adequate indexes. Table 13.12 lists some useful ones.

At a parochial level, materials suppliers will be listed, and may be categorised, in directories from Local Chambers of Commerce and of Trade, and also in some of the more extensive town directories, that from Milton Keynes being a good example.

County Authorities, for example Essex, sometimes sell directories of their industrial estates, which makes the physical location of suppliers easy.

A useful directory of US manufacturers is the *US Industrial Directory* produced by Cahners Publications. Even more comprehensive is the *Thomas Register of American Manufacturers*, published by Thomas Publishing Company. The 18 volumes of the 7th Edition (1984) comprise ten listing products and services, two providing company profiles and six reproducing company product catalogues.

CBD (1981) provide a guide to national, city and specialised directories and similar books for all European Countries. The same publisher also offers other similar publications for Africa, and for Asia and Australasia. Croner Publications of Queens Village, New York, USA, more ambitiously publish a regularly updated loose-leaf binder entitled *Trade Directories of the World*.

Some periodicals publish annual directories of services and suppliers; examples are *Wire Journal* and *Metals Finishing*.

14

Standards and Materials

When selecting materials the designer must be aware of those quasi-legal parameters against which his choice will be evaluated. He will initially generate a specification for his product: this will list comprehensively the technical and environmental requirements of the product for which materials must be found. Standards are then usually set up, defining technical performance limitations of candidate materials, processes and engineering practices, and will be used to validate final choices.

Specifications and standards are vital to successful industrial activity at all stages between product design and point of sale, where the customer's view of and expectations from his purchase have been moulded by what standards led him to expect.

14.1 Types of standards

Value standards are those which society expects to apply to the community as a whole: they include, for example, standards for clean water, regulation of emitted toxic gases and of radioactivity from industry. Davis (1984) has produced a review of UK environmental specifications in the last 25 years.

Regulatory standards may be developed by an industry to serve its own needs or image, or they may be imposed as mandatory by a Government Department. Examples are Health and Safety practices, and the control of industrial liquid effluent compositions.

Our present interest is specifically with materials and methods standards, which define achievable, and hence expected, performance parameters and the associated testing methods. The need for standards in industry was reflected by a London Conference on *The Role of Standards in the Industrial Strategy* (Department of Prices and Consumer Protection, 17th May 1978), and has since been re-affirmed many times.

14.2   Standards available

All civilised countries operate a national Standards activity, embodying locally accepted rules and recommendations for testing properties of materials and for defining those parameters which may be exploited in designs.

The award of a Standards compliance mark often confirms acceptability of a product in the market place, following the testing of materials or products in special test houses. Of these, the Underwriters' Laboratory in the USA is perhaps the most famous.

Standards most commonly cited in the materials world are those from the American Society for Testing Materials (ASTM), and the German DIN Standards. British Standards are listed in the annual *BSI Year Book*, and information about them can be obtained from BSI Headquarters in London or from the BSI Information Services and Marketing Department in Milton Keynes. Complete sets of British Standards are maintained in many UK and overseas libraries, listed in the current *BS Year Book*. A list of names and addresses of the leading National Standards organisations in the world is provided by Westbrook (1981).

Some materials, particularly metals, are defined by groups of British Standards. BS 1470, 1472, 1473, 1474, 1475, 1490 and 2901 for example embrace various aspects of aluminium alloys, as ingots, castings and different geometrical forms, and as filler rods and wires for welding. BS 1161 deals with aluminium alloy sections for structural purposes and other standards cover the use of aluminium in windows, lighting columns and building applications. The situation is even more complex for steels. To assist the student, the British Standards Institute offers Sectional Lists, which collect together all the related British Standards on a topic. For example, *Building* is covered by SL16, *Iron and steel* by SL24, and *Leather, plastics and rubber* by SL12.

The BSI Standing Order scheme periodically issues collated details of the various standards on plastics and rubbers. Sometimes commercial ventures offer collections of standards information on specified subjects; Brown (1981) presents test methods for plastics. Such collections may, of course, be unhelpfully biassed.

There is considerable activity at present by nations in trying to harmonise their Standards with the activities of the International Standards Organisation. Many British Standards are currently dual-numbered, bearing an ISO number and a British number. Unfortunately, experience has shown that International Standards take even longer to produce than do National ones, so the reader may best be advised to depend primarily upon his National Standards.

American standards are a special case. More than 420 different organisations in America develop standards for various areas of use, there being no central Federal Standardisation body. A valuable source of information on American

Standards is Technical Standards Services Ltd. of Aylesbury, Bucks; this Company specialises in procurement of American standards for the British customer.

14.3 Codes of Practice

Mention must be made, in this context, of Codes of Practice and of Safety Requirements. Codes of Practice apply in most countries. Purely as an example, BS CP 118 1969 *Structural Use of Aluminium*, defines permissible design stresses for various modes of mechanical loading on components made of nine alloy types. It also includes the intrinsic fatigue strength of the metal and of welded and riveted joints. Again, BS CP 117 Parts 1 and 2 deal with composite constructions based on steel. An overseas example is the American ANSI/ASME Boiler and Pressure Vessel Code, of which Section 2 covers materials specifications for ferrous and non-ferrous metals, welding rods, electrodes and fillers, and Section 10 deals with fibre-glass reinforced plastic pressure vessels. Information on American Codes of Practice is available from American Technical Publishers Ltd, Hitchin, Herts, while in the UK, Codes of Practice are obtainable from the British Standards Institution, London.

Sometimes, Codes of Practice are enforced by Government Agencies. The Health and Safety Executive in the UK is responsible for industrial safety and prescribes many material handling codes which might affect a designer's choice of materials. Purely as an example, the *Health and Safety Executive Report No.4, HMSO 1975*, deals with health precautions to be adopted by industry in handling, cutting and drilling asbestos-cement. Further information on all the Codes of Practice can be obtained from the Health and Safety Executive in Sheffield.

14.4 References on standards

Some reference sources on Standards should be cited. English translations of German Standards are listed in an *Annual Catalogue* with subject index, published by Deutsches Institut for Normung ev. Berlin, West Germany.

The International Standards Organisation Annual, with subject index, is published by ISO, Geneva, Switzerland. A world-wide microfilm and microfiche listing of many Standards by title and reference number, with an extensive materials subject index in hard copy, is available as *Industry Standards*, from Information Handling Services, Englewood, Col. USA.

The International Technical Information Institute of Japan publishes a *World Standards Material Speedy Finder*: Volume 4 *Materials* provides equivalents for materials described in US, UK, German, French, Japanese and ISO Standards.

The American Society for Testing and Materials produces an *Annual Book of ASTM Standards*, available through American Technical Publishers Ltd, Hitchin, Herts. Fifteen of the 66 volumes of current ASTM Standards refer to materials, and are grouped together for user convenience. Thus Volume 01.06 covers *Coated Steel Products*, Volume 08.04 covers *Plastic Pipe and Building Products*, and so on.

References

Al-Hassani & Sharkery M (1980). 'Powder metallurgy compacts by electrical discharge'. *Metal Engineering,* April, p.68.

Alexander, W.O. (1979). 'Designing to conserve energy', *Engineering,* December, p.1560.

Anderson, E.W. (1983). *'The European Aerospace Manufacturing Industry: The Supply of Strategic Materials'.* Report to the Ministry of Defence-Research (Materials and Collaboration) Department. December.

Anderson, I.G. (1985). (Ed.) *'Current British Directories',* Beckenham CBD Research Ltd.

Anon. (1979). 'Cellular plastics for building'. *Building Research Establishment Digest,* April, pp.1-7.

Anon. (1980a). 'A new way of selecting materials', *Design Engineering,* 51 (7), July, p.31.

Anon. (1980b). 'Non-black reinforcing filler for elastomers', *Design Engineering,* July, 18.

Anon. (1980c). 'Fibreglass spring', *Machine Design,* 52 (5), March, p.20.

Anon. (1980d). 'First nylon elastomer', *Design Engineering,* May, p.32.

Anon. (1981a). 'Energy conservation in injection moulding', *Technocrat,* 14 (6) June, p.43.

Anon. (1981b). 'Dispersion strengthened copper', *Electronic Design,* 29 (10), May, p.47.

Anon. (1981c). 'Explosive forming for rear axles', *Iron Age Metalworking International,* 20, (2), p.9.

Anon. (1981d). *'Who's-Who of British Scientists',* Dorking, Simon Books.

Anon. (1981e). 'Materials for electronic, industrial and newer ceramics', *Ceramic Industry,* 16 (1), January, pp.25-47.

Anon. (1982). 'Ionic nitriding process optimises diffusion', *Iron Age Metalworking International,* 6, June, p.23.

Anon. (1983a). 'Materials, metals, plastics fundamentals', *Electrionics,* 29 (4), April, pp.59-109.

Anon. (1983b). 'New material for metal cutting', *Tooling,* 37 (12), December, pp.24-28.

Anon. (1983c). 'Operating temperature is a vital statistic', *Eureka,* 3 April, p.39.

Anon. (1983d). 'Diffusion bonding', *Engineer,* 256 (6638), June, pp.32-33.

Anon. (1983e). 'Toxic fumes killed passengers in DC-9', *New Scientist,* 23 June, p.839.

Anon. (1984a). 'All the designer needs to know about engineering thermoplastics', *Eureka,* 4 June, p.15.

Anon. (1984b). 'High speed water to cut glass', *Engineer,* 258 (6669), January, p.37.

Anon. (1984c). 'Beamed heat cuts superplastic forming costs', *Machine Design,* 56 (2), January, p.8.

Anon. (1985). 'Light-weight composite tow bar', *NTIS Report AD-A 130*, 384, USA.

ARCO Metals Company (1983). 'Largest composite billet', *Machine Design*, 55 (25), November, pp.14-15.

ASLIB (1981). *UK Online Search Services*, London ASLIB.

Baker, S. (1980). 'Detecting fine cracks in ceramic substrates', *Electronics Times*, 30 April, p.14.

Baxter, C.P.G & Booth, G.S. (1979). 'Improved fatigue strength in welds', *Welding Institute Research Bulletin*, 20 (9), September, p.257.

Beard, D.R. (1980). 'Assessing casting quality', *Engineering*, November, p.1247.

Beck, R.D. (1980). *'Plastic Product Design'*, New York. Van Nostrand Reinhold.

Bishop, R. (1982). 'Most designers have a lot to learn', *Eureka*, 2 July, p.3.

Bittence, J.C. (1983). 'When computers select materials', *Materials Engineering*, 97 (1), January, pp.38-42.

BNCM (1984). 'Guide to the selection of plastics materials', *Designer's Guide to Materials*, Section 4.5, Harlow, British National Committee on Materials.

BPR (1983). 'It hasn't all been discovered yet', *British Plastics and Rubber*, September, pp.55-62.

Brooks, A. (1983). 'Mechanical plating', *Metal Finishing*, August, pp.53-57.

Brown, R.P. (1981). *'Handbook of Plastics Test Methods'*, 2nd Edn, London, G. Godwin.

Calayag, T. & Ferres, D. (1982). 'Zinc plus aluminium equal new bearing opportunities', *Automotive Engineering*, 90 (9), September, pp.40-44.

CBD (1981). *'Current European Directories'*, 2nd Ed. Beckenham. CBD Research Ltd.

Chadwick, G. (1984). Unpublished communication.

Chapman, A.H. (1984). 'Tin and tin alloy coatings - a review', *Leeds Materials Engineering Conference Proceedings*, pp.159-166. London. Institution of Metallurgists.

Clarke, G. (1981). 'Vitrous enamel - a quality finish', *Product Finishing*, 34, (1), January, p.30.

Clarke, T.C. (1980). 'Controlled doping of polyacetylene', *IBM Technical Disclosure Bulletin*, 23 (2), July, p.772.

Coates, P.D. & Ward, I.M. (1981). 'Die drawing: solid phase drawing of polymers through a converging die', *Polymer Engineering and Science*, 21 (10), July, pp.612-618.

Codlin, E.M. (1982). *'Directory of Information Sources in the UK'*, 5th Edn, Vol.1, London ASLIB.

Cogswell, F.N. & Hasler, R. (1984). 'Aromatic polymer composites', *Leeds Materials Engineering Conference Proceedings*, pp.141-145. London. Institution of Metallurgists.

Colbert, S. (1980). 'Designing with structural foam thermoplastics', *Electronic packaging and production*, 20 (6), June, p.195.

Collins, J.A. (1981). *'Failure of Materials in Mechanical Design'*. Chichester, J. Wiley.

Constable, G. (1980). 'Making the most of materials', *Engineering*, November, p.1234.

Constance, J.A. (1981). 'Gas gun produces net shapes from p/m', *American Metal Market*, 89 (114), June, p.12.

Cooper, E. (1982). *'New Trade Names in the Rubber and Plastics Industries'*, Shawbury. RAPRA Technology Ltd.

Das, A.A.& Chatterjee, S. (1981). 'Squeeze casting of aluminium alloy', *Metallurgist and Materials Technologist*, 13 (3), March, p.137.

Davidge, R.W. (1983). 'Zirconia toughening of ceramics', *Materials Development News*, (UKAEA, Harwell) (26) November, p.1.

Davis, K.C. (1984). 'An overview of environmental test specifications 1958-1983', *Journal of the Society of Environmental Engineers*, March, pp.23-30.

Dhingra, A.K. (1981). 'Fibres in metal castings', *Chemtech*, 11 (10), October, p.600.

Diegle, R.B. (1981). 'Low cost glass coatings', *Materials Engineering*, 96 (1), July pp.46-49.

Dieter, G.E. (1983). *'Engineering Design – a Materials and Processing Approach'*, New York. McGraw Hill.

Doane, D.V. (1984). 'Materials - Whose responsibility?' Editorial. *Molybdenum Mosaic – The Journal of Molybdenum Technology*, 7 (2).

Drake, M.P. (1980). 'Electrodeposition in ultrasonic fields', *Institute of Metal Finishing Transactions*, 58 (2), p.67.

Duckworth, W.F. (1984). 'The challenge to the materials technologist', *Materials and Design*, 4, December/January, pp.924-926.

Dumper, L.J. & Loader J.M (1984). 'How do I find out about..? No.12 - Where to find out.' *Welding Institute Research Bulletin*, 25 (12) December pp.420-427.

El-Sherbiny, M. & Salem, F. (1981). 'Surface protection by ion plated coatings', *Anti-Corrosion*, 28 (11), November, p.15.

Fink, D.E. (1980). 'Availability of strategic materials debated', *Aviation Week and Space Technology*, 5 May, pp.42-46.

Ghouse, M. (1980). 'Occlusion plating of copper-silicon carbide composites', *Metal Finishing*, 78 (3), March, p.31.

Gill, B.J. (1984). 'Designing and producing engineering surfaces', *Leeds Materials Engineering Conference Proceedings*, 191-208. London. Institution of Metallurgists.

Grace, K. (1984). 'Metallised plastic components', *British Plastics and Rubber*, April, p.25.

Graham, J.W. (1981). 'What can designers do for business? Ask top management'. *Industrial Design*, May/June, pp.24-26.

Gray, A.G. (1980). 'Materials substitution: challenge of the 80s', *Metal Progress*, 117 (2), February, pp.32-37.

Grayson, S.J. Hume, J. & Smith, D.A. (1982). 'Reduction of smoke and toxic gases from flexible polyurethane foams under fire conditions: introductory review and chemical model', *Plastics and Rubber Processing and Applications*, 2, pp.111-122.

Hall, F. (1981). 'Value engineering in plastics design', *Plastics Engineering*, 37 (2), February, pp.23-33.

Hall, R.O. (1979). 'Rubber as an engineering material', *Materials in Engineering Applications*, 1, September, p.295.

Hamme, J.V. (1979). 'Ceramics (Raw Materials)', *Kirk-Othmer Encyclopaedia of Chemical Technology*, 3rd Edn. Vol.5, New York, J.Wiley.

Harrison, J.D. (1980). 'Lessons from service fatigue failures', *Welding Institute Research Bulletin*, 21 (3), March, p.68.

Hartley, P. (1980). 'Plastics assembly methods', *Engineering*, December, p.1356.

Harvey, A.P. & Pernet, A. (1981). *'European Sources of Scientific and Technical Information'*, Harlow, Longmans.

Haugen, E.B. (1982a). 'Modern statistical materials selection Part 1', *Materials Engineering*, 96 (12), July, pp.21-25.

Haugen, E.B. (1982b). 'Building a data base', *Materials Engineering*, October, pp.67-70.

Haugen, E.B. (1982c). 'The price of safety', *Materials Engineering*, September, pp.49-51.

Haugen, E.B. (1982d). 'Random variables and reliability', *Materials Engineering*, 96, August, pp.49-51.

Haugen, E.B. (1982e). 'Correlation and computers', *Materials Engineering*, 96, November, pp.65-68.

Hegin Galvano Aluminium (1982). 'Electrodeposition of aluminium', *Metal Finishing Plant and Processes,* 18 (1), p.27.

Henderson, G.B. & Henderson, S.P.A. (1980). (Eds.) *'Directory of British Associations',* 6th Edn. Beckenham, CBD Research Ltd.

Hinds, L. (1983-4). 'Rapidly growing uses for expanded PTFE', *Plastics Today,* (18), Winter, pp.16-19.

Hioki, S. (1980). 'Nitriding of aluminium by surface melting', *Journal of Heat Treating,* 1 (3), June, p.65.

Hodesblatt, S. (1982). 'Alternatives to strategic materials', *Design Engineering,* January, pp.44-49.

ICI (1984). 'Tailor-made materials for engineering applications', *Engineering Plastics,* (19), Summer, p.2.

Jack, K.H. (1984). 'Engineering applications of sialons', *Leeds Materials Engineering Conference Proceedings,* pp.65-69. London. Institution of Metallurgists.

Jacobson, D.M. & Evans D.S. (1983). 'The supply economics of niobium', *The Metallurgist and Materials Technologist,* 15 (11), November, pp.566-567.

Jenkin, W.C. (1980). 'A new forming process', *Manufacturing Engineering,* June, p.108.

John, C.G. (1984). 'Electroplated coatings for wear and product improvement', *Leeds Materials Engineering Conference Proceedings,* pp.183-190, London. Institution of Metallurgists.

John, S. Perumal, A. & Shenoi, B.A. (1984). 'Chemical colouring of aluminium', *Surface Technology,* 22, pp.15-20.

Judd, N.C.W. (1983). Personal communication.

Judd, N.C.W. (1984a). Personal communication.

Judd, N.C.W. (1984b). Personal communication.

Kaelble, D.H. (1983). *CAD/CAM Handbook for Polymer Composite Reliability,* Vol.1, NTIS Report AD-127, p.720, USA.

Kerr, J. (1981). 'Glassy metals in transformers', *Engineer,* 252 (6517), February, p.33.

Knight, R. (1982). 'Superplastic stainless slashes tool costs', *Eureka,* 2 September, pp.85-86.

Knight, R. (1983a). 'Britain gets lithium super-alloys off the ground', *Eureka,* 3 September, pp.30-31.

Knight, R. (1983b). 'Plastics - the popular choice for production', *Machinery and Production Engineering,* November, pp.18-22.

Knight, R. (1983c). 'Operating temperature is a vital statistic', *Eureka,* 3 April, pp.39-44.

Knight, R. (1983d). 'Fibre in GRP foam sandwiches', *Eureka,* 3 (1), January, p.25.

Koenig, C.L. & Yust, C.S. (1981). 'Explosive Compaction of AlN, amorphous Si_3N_4, boron and Al_2O_3 ceramics', *American Ceramic Society Bulletin,* 60 (11), November, p.1175.

Koplev, A.Lystrup, A. & Vorm, T. (1983). 'The cutting process, chips and cutting forces in machining CFRP', *Composites,* 14 (4), October, pp.371-376.

Lamberson, L. (1981). 'Materials reliability', in *Kirk-Othmer Encyclopaedia of Chemical Technology,* 3rd Edn, Vol.5, New York, J.Wiley.

Leslie, P.J. (1982). 'The role of industrial design', *Electrical Communication,* 57 (1), pp.4-6.

Liddiard, E.A.G. (1980). *'Register of Consulting Scientists, Contract Research Organisations and Other Scientific and Technical Services',* Slough, Fulmer Research Institute.

Lingenfelder, H. (1983). *'World Guide to Special Libraries'*, Munich, K.G. Sauer.

Machining Data Handbook (1980). *'Machining Data Handbook'*, 3rd Edn, Ohio, Metcut Research Association Inc.

Materials Engineering. (1982). 1983 Materials Selector. *Materials Engineering*, 96 (6).

Matthews, A. (1984). 'The value of deposition processes for industrial tools', *Leeds Materials Engineering Conference Proceedings*, pp.175-182. London. Institute of Metallurgists.

Metal Finishing (1984). *'Metal Finishing'*, New Jersey, Metals and Plastics Publications Inc.

McCafferty, E.Shafrin, E.G. & McKay, J.A. (1981). 'Microstructure and surface modification of an aluminium alloy by rapid solidification with a pulsed laser'. *Surface Technology*, 14, November, pp.219-223.

McIntyre, R.D. (1982). 'ODS alloys', *Materials Engineering*, 95 (2), February, pp.34-39.

Mesker, P.J. (1978). 'Does anyone care about the energy used in materials and manufacturing?' *Materials Engineering*, September, pp.37-40.

Michael, A.D. & Hart, W.B. (1980). 'SME brass - a new memory metal', *The Metallurgist and Materials Technologist*, 12 (8), August, p.434.

Millard, P. (1971). *'Trade Associations and Professional Bodies in the UK'*, Oxford, Pergamon Press.

Miller, K.J. (1980). 'Educating the design engineer in materials technology', *Metals and Materials*, June, p.16.

Miska, H.A. (1980). 'Chemically strengthened glass', *Design Engineering*, March, p.57.

Mitchell, E.& Lynes, E. (1983). *'Materials Research Centres'*, Harlow, Longmans Group.

Moakes, R.C.W.& Norman, R.H. (1977). 'Techniques and equipment for assessing the environmental behaviour of non-metallic materials', *Engineers' Conference*, Ware. Society of Environmental Engineers.

Moritz, J.A. (1980). 'Quality control for injection moulding', *Plastics World*, September, p.60.

Morrell, R. (1980). 'Engineering ceramics - property data for the user', *Materials in Engineering*, 2, September, pp.15-20.

Newnham, R.C.& Davies, D.G.S. (1976). 'Comparison of engineering properties of ceramics', *Materials Optimizer*, Section 3/C/4, Slough, Fulmer Research Institute.

Norman, J.C. (1981). 'Synthetic fibres find applications in transportation', *Industrial Research and Development*, July, pp.105-108.

Oakley, M. (1984). 'Managing design'. *Engineering*. November, pp.800-801.

Pastorini, N. (1984). Editorial, *Engineering Design*, (DuPont), 2.

Phelan, D.& Wyatt, L.M. (1981). 'Manufacture of metal components', *Materials Optimiser*, Section 1D-M, Slough, Fulmer Research Institute.

Piazza, S. (1981). 'An overview of thermoplastic rubbers', *European Rubber Journal*, November 9.

Pilling, S. (1984). (Ed.) *'Research in British Universities, Polytechnics and Colleges'*, Wetherby, British Library.

Pope, J.R. (1980). 'The importance of sound mechanical design', *Engineering*, November, p.1238.

Pye, A.M. (1979). 'Composite producing techniques', *Materials Optimiser*, Section 1D, Slough, Fulmer Research Institute.

Pye, A.M. (1982). 'Transparent polymers offer optical options', *Design Engineering*, March, pp.58-63.

Pye, A.M. (1983*a*). 'Metal-like polymers', *Design Engineering*, August, pp.38-39.

Pye, A.M. (1983*b*). 'Computer usage in materials selection and engineering design', *Design Engineering*, October, pp.129-136.

Quinn, J.A. (1977). *'Design Data—Fibre Glass Composites'*, St. Helens, Fibreglass Ltd.

Rams, D. (1983*a*). 'Industrial design in a time of change', *Materials and Design*, 4, April/May, pp.706-709.

Rams, D. (1983*b*). 'The designer's contribution to company success', *Materials and Design*, 4, June/July, pp.771-775.

Reid, C.N.& Greenberg, J. (1980). 'An exercise in materials selection', *Metallurgist and Materials Technologist*, 12 (7), July, p.385.

Satoh, S.(1981). 'Nitriding process increases production', *Iron Age Metalworking International*, 20 (8), September, p.41.

Seah, M.P. Lea, C. & Hondros, E.D. (1981). 'The fragility index for quality: intergranular failures', *Metallurgist and Materials Technologist*, 13 (11), November, p.555.

Smallen, M. (1981). 'Weld cracks from braze material', *Electronic Packaging and Production*, 20 (11), November, p.136.

Smith, C.A. (1981). 'The microbiology of corrosion', *Anti-corrosion*, 28 (1), January, p.15.

Smith, C.A. (1982). 'Corrosion of metals by wood', *Anti-corrosion*, 29 (3), March, p.16.

Smith, T. (1980). 'Ten ways to reduce corrosion', *Anti-corrosion*, 27 (12), December, p.13.

Sprecher, N. (1984). 'New applications inherit high-performance composites', *Design, Products and Applications*, 6 (12), December, p.18.

Stasko, D. (1980). 'Machineability characteristics of aluminium', *Modern Casting*, 70 (4), April, p.74.

Stone, C.R. & Campbell, J.M. (1984). 'New approaches to the use of short fibre rubber composites in engineering applications', *Leeds Materials Engineering Conference Proceedings*, London. Institution of Metallurgists.

Thorne, A. (1980). *Thermoplastic Elastomers*, Shawbury, RAPRA Technology Ltd.

Turner, L.W. (1979). 'Processing techniques for polymers', *Materials Optimiser*, Section ID-P, Slough, Fulmer Research Institute.

'Ulrich' (1984). *'International Periodicals Directory'*, 23rd Edn, New York, RI Bowker Co.

US Department of the Army. (1981). *'Discontinuous Fibre Glass Resin-Faced Thermoplastics'*, Engineering Design Handbook, DARCOM Pamphlet No. 706-314, April.

van de Berg, J.F.M.Daerman, T.E.G., Krijl, G. & van de Leest, R.E. (1980). 'The electrodeposition of aluminium', *Philips Technical Review*, 39, (3/4), pp.87-91.

Varjian, R.D. & Hall, D.E.C. (1984). 'Report of the electrolytic industries for the year 1983', *Journal of the Electrochemical Society, Reviews and News*, 131 (9), September, pp.374C-400C.

Vaughn, T.B. (1981). 'Powder enamelling' *Product Finishing*, 34 (1), January, p.37.

Walshe, K.B.A & Lowe, P.H. (1984). 'Computer-aided cost estimation for injection moulds' *Plastics and Rubber International*, 9 (4), August, pp.30-35.

Wang, N.M. & Ayres, R.A. (1980). 'CAD for sheet metal', *Materials Engineering*, 93, November, p.52.

Waterman, N.A. & Pye, A.M. (1980). 'Guide to the selection of materials for marine applications', *Materials Optimiser*, Section 1V-M2-d. Slough, Fulmer Research Institute.

Waterman, N.A. (1979). 'Product failure due to bad material choice and processing', *The Implications of Materials Selection on Product Liability Conference Proceedings*, pp.18-27, London, Materials in Engineering.

Waterman, N.A. (1984). 'Materials for the 1980s and 1990s', *Metallurgist and Materials Technologist,* 16 (9), September, pp.461-463.

Watson, P. (1979). 'Materials selection for durability', *The Implications of Materials Selection on Product Reliability Conference Proceedings,* pp.59-69, London, Materials in Engineering.

West, G.H. (1984). 'An approach to process economics in the plastics industry', *Plastic and Rubber International,* 9 (1), February, pp.34-35.

Westbrook, J.H. (1981). Materials Standards and Specifications, *Kirk-Othmer Encyclopaedia of Chemical Technology.* 3rd Edn, Vol.15, pp.35-56, New York, J. Wiley.

Whelan, T. (1981). 'Transparency choice' *British Plastics and Rubber,* May, pp.40-41.

White, P.E. (1984). 'PTFE coatings for wet environments', *Leeds Materials Engineering Conference Proceedings,* pp.221-225, London, Institution of Metallurgists.

Williams, T.I. (1980). (Ed.) *'Industrial Research in the UK',* Harlow, Longmans Group.

Williams, T.I. (1982). (Ed.) *'European Research Centres',* Harlow. Longmans Group.

Wilms, G.R. & Aghan, R.L. (1981). 'Anisotropy in machining of steel plates', *Metals Technology,* 8 (3), March, p.108.

Wimberly, P.A. (1985). 'Speeding engineer access to technical information', *Machine Design,* 57 (2), January 24, pp.86-90.

Young, R.J. (1984). 'Conducting polymers - plastic metals? *Plastics and Rubber International,* 9 (1), February, pp.29-33.

Zweben, C. (1984). *Introduction to the Principles of Metal Matrix Composites.* Short course on 'The technology of fibre-reinforced composite materials: properties, processing and evaluation', 16-20 July. University of Surrey.

Bibliography

In addition to the quoted references which are specific to individual chapters and subjects, many recent books on materials subjects are available. A selection of these is listed, alphabetically by author, as an aid to further study, several having already been mentioned in the previous text.

ASTM (1981).'*Fatigue of Fibrous Composite Materials*'. STP 723. Philadelphia. ASTM.

ASTM (1982).'*Fire Test Standards—a Compilation of 71 Standards*'. Philadelphia. ASTM.

ASTM (1983) '*Behaviour of Polymeric Materials in Fire*'. Philadelphia. Special Technical Publication STP816. ASTM.

Ashby, M.F. & Jones, D.R. (1979). '*Engineering Materials: an Introduction to their Properties and Applications*'. Oxford. Pergamon Press.

Bainbridge, C.G. (1980). '*Welding*', 4th Edn Sevenoaks. Hodder and Stoughton.

Beck, R.D. (1980). '*Plastic Product Design*'. London. Van Nostrand Reinhold Co.

Boyer, H.E. & Gall, T.L. (Eds.). (1985). '*Metals Handbook, Desk Edition*'. Ohio. American Society of Metals.

Brown, R.P. (1981). '*Handbook of Plastics Test Methods*' 2nd Edn, London. George Godwin Ltd.

Chatfield, C. (1983). '*Statistics for Technology*'. London. Chapman & Hall.

Collins, J.A. (1981). '*Failure of Materials in Mechanical Design*'. New York. J. Wiley & Sons.

Cook, N.H. (1984). '*Mechanics and Materials for Design*'. London. McGraw, Hill.

Creyke, A. Sainsbury, D, & Morrell, G. (1982). '*Design with Non-Ductile Materials*'. London. Applied Science.

Dodd, A.E. (1968). '*Dictionary of Ceramics*'. London. G.Newnes.

Dyan, J.B. (1983). '*Product Design with Plastics*'. New York. Industrial Press Inc.

Farag, M.M. (1979). '*Materials and Process Selection in Engineering*'. London. Applied Science.

Fellowship of Engineering (1983). '*Modern Materials in Manufacturing Industry*'. London. FOE.

Hertzberg, R.W. (1983). '*Deformation and Fracture Mechanics of Engineering Materials*'. 2nd Edn, London. J. Wiley.

Howard, M.J. (Ed.) (1980). '*Plastics—a Desk-top Data Bank*'. San Diego. The International Plastics Selector Inc.

Howard, M.J. (Ed.) (1977). *'Elastomeric Materials – a Desk-top Data Bank'*. San Diego. The International Plastics Selector Inc.

Howard, M.J. (Ed.) (1978). *'Foams – a Desk-top Data Bank'*. San Diego. The International Plastics Selector Inc.

Jeffries, J.L. (Ed.) (1982). *'Materials, Metals, Plastics and Parts Glossary'*. Insulation/Circuits, April, p.57-86.

Lockett, F.J. (1982). *'Engineering Design Basis for Plastics Products'*. London HMSO.

Lovelock, D.W. & Gilbert, R.J. (1975). *'Microbiological Aspects of the Deterioration of Materials'*. London. Academic Press.

Lowenheim, F.A. (1982). *'Guide to the Selection and use of Electroplated and Related Finishes'*. STP 785. Philadelphia. ASTM.

Lubin, G. (1982). *'Handbook of Composites'*. London. Van Nostrand Reinhold.

McColm, I.J. (1983). *'Ceramic Science for Materials Technologists'*. Glasgow. Leonard Hill.

Mills, J.A. & Redford, D (1983). *'Machineability of Engineering Materials'*. London. Applied Science.

Moroney, M.J. (1956). *'Facts from figures'* London. Penguin.

Quenoille, M.H (1959). *'Rapid statistical calculations'*. London. Chas Griffin.

Rettit, T. (1981). *'Appreciation of Materials and Design'*. Philadelphia. International Ideas Inc.

Richardson, D.W. (1982). *'Modern Ceramic Engineering'*. New York. Marcel Dekker Inc.

Ross, R.B.(1977). *'Handbook of Metal Treatments and Testing'*. London. E & FN Spon.

Shackelford, J.F. (1984). *'Introduction to Materials Science for Engineers'*. London. Macmillan.

Sheldon, R.P. (1982). *'Composite Polymeric Materials'*. London. Applied Science Publishers.

Simons, E.N. (1969). *'A Dictionary of Alloys'*. London. F. Muller.

Simons, E.N. (1970). *'A Dictionary of Ferrous Metals'*. London. F. Muller.

Smithells, C.J. (1983). *'Metals Reference Handbook'*. 6th Edn, London. Butterworths.

Struik, L.C.E. (1978). *'Physical Ageing in Amorphous Polymers and other Materials'*. Oxford. Elsevier Scientific Publishing Co.

Unterweiser, P.M. (1979). *'Case Histories in Failure Analysis'*. Metals Park, Ohio. American Society for Metals.

Walker, B.M. (Ed.) (1979). *'Handbook of Thermoplastic Elastomers'*. London. Van Nostrand Reinhold Co.

Wordingham, J.A. & Rebone, P. (1968). *'Dictionary of Plastics'*. London. Newnes Books.

Various authors. *'Materials Optimiser 1974'*. Slough. Fulmer Research Institute.

Index

abrasion, *see* wear
accelerated testing, *see* testing of materials
adhesion, *see* plastics
  bond strength 221, 222
  joint design 222
  materials 222
aesthetics 12, 182, 187
alloys, *see* individual metals
aluminium
  alloys 58, 69
  hardening 69
  sources 55
  surface protection 69, 70
  world production 55
annealing of metals 62
anodic coatings, *see* coatings
  for aluminium 195, 201
arc
  electric, for welding 232
  plasma 234

bending properties, *see* fatigue
  metals 40, 61
bibliography 272
biological agents
  see corrosion
  see environmental vectors
  attack on materials 147, 153, 164
brazing, *see* joining processes
  alloys 37
  processes 36, 90

brittleness, *see* fracture
  ceramics 79, 82, 149
  plastics 100, 152
burning rate, *see* flammability

carbides, *see* ion bombardment
  cutting tools 150
casting, *see* encapsulation methods
  metals 42, 230, 233
  processes 57, 64, 230, 233
  resins 230, 237

ceramics, *see* coatings, ceramic
  bonding 79
  ceramic fibre paper 87
  composites 81, 84
  design factors 85, 90
  effects of environment 82, 150
  impact resistance 85, 149
  manufacturing processes 48, 79, 82, 85, 86, 88, 89, 238
  materials and their varieties 80
  products 86, 87, 239
  properties of 48, 79, 83
  refractories 81
  sialon 85
  silicon carbide 80, 86
  silicon nitride 86
  standard products 81
  structure 79, 80
  wear resistance 82, 149
  zirconium oxide 86

chromium, *see* surface finishing
  alloys 56
  coatings 194, 195, 222
  in conversion coatings 201 ·
  oxide-based coatings 196, 201
  pack chromizing 189, 203
  supplies 55
  world sources 55, 225
cleaning
  before surface finishing 182
  of metals 182, 191
coatings
  see electroplating
  see painting
  see plasma
  see surface finishing
  aesthetics 182, 184, 188
  anodic 195, 201
  ceramic 195, 204
  conversion 182, 189, 196, 200
  diffusion 194, 203
  hot dip 188, 203
  metal 188, 189, 195
  paint 183, 187, 191, 205
  plastic 189, 192
  spray 190, 197
  vapour deposition 190, 201
  vitreous 204
codes of practice, *see* standards
cold working
  of metals 34, 39, 45, 61, 64
  processes, *see* welding
    forging 231, 233
    forming 231
    pressing 231
    rolling 231
    stamping 231
compacts
  of ceramic 238
  of metal powder 231, 232
composites
  anisotropy 48, 135
  ceramics 124
  constructions 124, 133
  creep in 48
  design data 133, 134
  fibres 43, 125, 127, 134
  fillers 125, 128

high strength 127, 135
inter-facial bonding 130, 134
limitations of properties 129, 133
manufacturing technology 130, 135, 137,
  237
metal-based 72, 124, 141, 163
plastic-based 99, 130, 139
properties 129, 131
reinforcements 43, 127
structures from 129, 151
varieties of 126, 130
weathering 133
computer
  cad 14, 212, 231
  cam 212, 231
  data bases 31, 247
  for materials selection 32, 247
concrete
  high strength 88
copper
  alloys 72
  hardening 72
  metal 71
  plating 187, 198, 200
  suppliers 55
  world sources 55
corporate image 8, 12, 220
  aesthetics 12
  design of products 8, 12, 220
corrosion
  atmospheric 58, 62, 146, 183
  biological 147
  chemicals 68, 194
  galvanic 68, 71, 147, 183, 214
  physical by cavitation 147
  prevention 62, 169
  products 58, 146
  protection from 58, 147, 187
  resistance 58
  tests 146
cost, *see* metals
    see plastics
  competitiveness 22, 54, 224
  effectiveness 21, 24
  of materials 27, 97
  of processes 25, 37, 57
creep
  effects 216

creep (cont)
  in service 144, 152
  limits of 30
  of materials 30, 84
  rates 216
  recovery 216
  rupture 71
  tests 32
crystallinity
  of ceramics 38, 79, 84
  of glasses 38
  of metals 38
  of plastics 5, 38
cutting, *see* carbides
        see machining
  by laser beam 234, 236
  by water jets 221, 239
  fluids 77
  metals 74, 77, 78
  plastics 114
  processes 74, 221
  rates 74, 76, 77, 78
  tools 75, 238

data, *see* information sources
  in books 241
  in computers 31, 33
  in the memory 31
design, *see* ceramics
        see composites
        see human factors
        see value engineering
  and profitability 2, 6, 8, 25
  brief from client 15, 18
  check-list for plastics 120
  criteria to be satisfied 13, 29, 208
  impact of ambient environment 147
  mechanical 21
  reviews 10
  simplification 22
  specifications 2, 6, 10,19
designer, *see* data
        see industrial design
  and R & D 18
  education of 1, 12, 17, 19, 120, 241
  intuition 6, 24
  materials selection by 20, 24, 27, 28, 31,
    119

social responsibility 10, 25
sources of materials information 19, 242,
    252
  use of consultants 8, 17, 18, 19, 28, 31,
    136, 252, 256, 259
ductility, *see* stress
  metals 145
  plastics 93, 100, 103, 106, 108

elastomers
  additives 95
  applications in products 96
  cellular foams 96
  coatings 95
  manufacturing methods 95
  properties 103, 104
  thermoplastics 96
  varieties of 96
electrical conduction
  of metals 43, 60, 72
  of plastics 122
  of solutions 198, 214
  semiconductivity 227
electroplating
  electroforming of artifacts 197, 199
  occlusion plating 199
  on metal substrates 195, 197
  on plastic substrates 195, 200
  processes 198, 239
  with metals 139, 187, 195
encapsulation methods
  coatings 70, 189, 207
  materials 192, 207
energy
  derived from materials 228
  effects of cost 41, 42
  locked into products 3, 41, 42, 57
  needs of processes 41
  specific, of materials 3, 42
environmental vectors, *see* weather
  acceleration of 143, 146, 208
  biological agents 147
  effects 155, 209
  test methods 109, 149, 208
failure
  affecting profits 212
  analysis 217
  lessons from 213

mechanisms 133, 210
  probability of 218
fatigue
  effects 84, 212
  metals, in vibration or bending 145, 209
  plastics 212, 216
  testing 212
  thermal 212
fibres
  as reinforcement 122, 128, 141, 151
  materials 49, 93, 126, 140
fillers
  in composites 126, 128
  in elastomers 95
  in plastics 92, 94, 139
fire
  avoidance of 169, 181
  causes 165, 167
  effects 169, 180
  testing methods 175, 177, 179
flame
  fire risk 48, 58
  spraying 197, 204
  use for depositing coatings 196

flammability
  and electric fields 170
  at elevated temperatures 173
  burning rate 171, 176
  combustion of materials 164
  −control additives in materials 165
  customer requirements 165, 166
  delayed ignition materials 168
  heat sources 170, 176
  layout of product to minimise 166, 171
  local requirements 164
  of components 168
  of materials 168, 172
  qualification testing of products 166, 180
  reducing in products 169
  standards 179
  test methods 173, 175, 177, 178
flaws *see* fracture
  consequences 48, 238
  detection 145, 215
flexibility
  elastomers 95

plastics 151, 153
forming processes
  ceramics 238
  composites 137, 138, 139, 237
  explosive 232, 238
  metals 230
  plastics 235
  powder routes 232
  work-hardening 39
fracture *see* flaws
  brittle 48, 61, 238
  ductile 41
  intergranular 39
  mechanics 84, 130, 143, 145, 215
  strength 24
  toughness 24, 29, 49, 61

galvanic effects
  corrosion 146, 183, 188
  series of metals 184, 214
glass
  borosilicate 81
  composition 81, 88
  manufacturing processes 81, 88, 89
  quartz 84, 85
  sol-gel process 81, 88
  structure 80
  testing 82
  uses 84

hardening processes
  ion bombardment 78, 202, 234
  metal 39, 58, 61, 67, 69, 72
    carburise 194, 234
    nitride 78, 190, 195
  precipitation hardening 45, 69
hardness
  metals 39, 41
  plastics 103
  processes to increase 39
  testing 41
  values for materials 47
hazardous materials
  asbestos 29, 222
  cadmium 222
  other 29
heat treatment
  see hardening processes

heat treatment (cont.)
  see laser beam
  see metals
  solution treatment of metal alloys 40
high temperature
  ceramics to withstand 43, 86, 150
  metals to withstand 62, 67, 148
hot working
  of metals 57, 68
human factors 8

impact *see* brittleness
  see ceramics
  strength 145
  testing 145
industrial design
  see design
  see designer
  concepts 9
  organisation in a company 15
information sources
  academe 252
  books 242, 245
  colleagues 241
  computer data bases 31, 247
  conferences 246
  consultants 8, 17, 19
  films 247
  government agencies 255
  industry 254, 256, 258, 259
  journals 242
  libraries 241
  patents 247
  professional institutes 256
  public corporations 256
  research associations 254
  suppliers 259
  trade associations 258
  trade papers 244
injection moulding
  costs-numbers trade-off 30
  metals (die casting) 21, 56, 57, 64, 71,
    233
  plastics 30, 114
  process affecting product 99
ion bombardment
  implantation in metals 78, 202
  plating 78, 202

nitriding 78, 202
iron *see* steel
  alloys 58, 67
  metal 67
  supplies 21

joining processes *see* soldering
                  see welding
  adhesives 236
  brazing 37, 73
  ceramics 79, 90
  mechanical fasteners 215
  riveting 236, 263

laminates 124, 139
laser beam
  cutting 236
  surface treatment of metals 234
  welding 236
levitation
  by acoustic waves 239

machining
  ceramics 48, 82, 90
  composites 237
  metals 56, 62, 74, 233
  plastics 114, 237
  relative indices of 75
  speeds and rates 76, 77, 78
  tools 76
  ultra-high speeds 233
magnesium
  alloys 59, 70
  corrosion 59, 70
  supplies 55, 70
  world sources 55, 70
magnetism
  devices 129, 217
  disadvantages 145
  materials 68, 73, 82, 129, 149
management
  expectations from designers 12
  of companies 13
  organisation of design function 8, 10, 14,
    17
  support for designers 15
manufacturing
  and relations with design 7

industry 7, 56
technology 24
market
  competitiveness 8
  forces 7, 8, 15
  requirements 2, 8
materials
  anisotropic 43, 48, 233
  comparison of families 43
  conservation 3, 42
  economics 22, 24, 27, 37, 97
  envelope of properties 39, 43, 48, 49
  forming processes 36, 39, 230
  isotropic 39
  publications of data 28, 31, 241
  ranking for usefulness 156
  reclamation 94
  recycling 3, 94
  selection 6, 20, 32
  shortages 27
  strategic 225
  substitution or alternatives 22
  suppliers 21, 259
  surface finish 20, 182
  testing 24, 30, 40
  varieties available 5, 27, 33
  waste in production 42, 56
material structures *see* ceramics
  engineering level 5, 38
  microstructural level 5, 38
  molecular level 5, 38
  testing 4
metals
  see cold working
  see corrosion
  see creep
  see fatigue
  see forming processes
  see fracture
  see heat treatment
  see weather
  see welding
  alloys 45, 57, 67, 69, 70, 71, 73
  costs 55, 61
  details of specific 28, 58,
  effects of environment 58
  fabrication methods 45, 56, 57, 230
  finishing 190

performance in service 144
properties 45, 46, 60
strength 41, 61, 141
substitution 22, 220
superplastic 69, 231
toughness 61, 142
work hardening 40, 61
world production 55
moulding *see* casting
        see injection moulding
  blow 97, 114
  extrusion 114
  green ceramics 89
  multi-functional parts 21, 48
  plastics 93
  processes 114

nickel
  alloys 73
  corrosion resistance 73
  supplies 73
  world sources 55, 226
non-metals *see* ceramics
  see composites
  see elastomers
  see films
  see glass
  see plastics
notch strength
  see impact
  see toughness
  crack propagation 2, 238
  sensitivity 145
  test methods 145

painting
  materials in paint 205, 207
  performance of paint 206
  processes 185
  spray painting materials 189
  spray painting processes 191, 192
  substrates for spray painting 182, 206
past experience
  of materials selected 29, 31
  rechecking 29
patents 247
physical properties
  of materials 35

physical properties (cont.)
  test methods 41
plasma
  cutting 234
  etching 75
  processing of surfaces 221, 233, 234
plastics
  see creep
  see elastomers
  see fatigue
  additives 92
  applications 95, 140, 151
  biological effects on 153
  cellular foams 96, 122, 141
  copolymers 101, 207
  costs relative to other materials 97
  degradation 100, 151
  fabrication-processing 94, 113, 235
  films 97, 151
  flammability 103, 168
  isotropy−anisotropy 99, 103, 105
  joining 117, 118, 119
  monomers 92, 114
  performance in service 93, 122, 151
  precursors 92
  properties 48, 99, 105
  pultrusion 130, 237
  resins 93, 139
  strain-dependence 45, 99
  structure 91
  thermoplasts 93
  thermosets 93
  trade names 33, 100, 105
  varieties of 93

pollution, *see* environmental vectors
  air 146, 180
  ground 57
  reclamation of materials 3, 42, 94
  water 223

polymers, *see* elastomers
  see plastics
porosity
  fired ceramics 89, 238
  plastic foams 96
powder, *see* compacts
  coating processes 192
  compaction to strip 232

metallurgy 231, 232
  use in cutting processes 239
pressure vessels 135, 263
processing
  affects product reliability 25
  of materials 221, 228, 239
  optimisation 4
product, *see* corporate image
  see cost
  see design
  see service life
  and its environment 2
  assembly 20, 36
  design of 19, 20
  external appearance 12, 13, 20
  life cycle 8, 209

quality
  auditing 216
  control methods 216
  definition 216
  evaluation 217
  product safety analysis 217
  sampling regimes 216
  testing 41, 217

raw materials
  availability 3
  costs 22, 24, 27
  restricted number 3
references 265
reliability
  definitions 210, 217
  forecasting 149, 208
  standards 263
  statistics 19
requirements for products 12
  customer expectations 143
  resistance to storage 144, 146
  service performance 143

safety, *see* quality
  and reliability 10, 15, 19, 29
  concepts 210
  derating factors 29, 107, 209
  factors 215, 218
  fire tests 164, 175, 177, 179
  of products 210

profile 217
statistics 218
service life, *see* plastics
   see product
   damage – tolerant criterion 211
   end of 209
   fail – safe criterion 210
   predictions 208
   safe – life criterion 210
shape-memory in metal alloys 72, 223
sheet drawing processes
   cold - 68, 69, 231
   deep - 68, 231
   for metal 58, 231
   warm - 70, 231
sintering
   ceramic powders 48, 238
   manufacturing processes 48, 238
   metal powders 222, 232
smoke emission,
   from burning materials 180
soldering
   materials 37, 224
   processes 37
solvents
   effects on plastics 109
specifications, *see* design
   customary in trade or industry 220, 261
   military 161, 163
   national 261
standards
   codes of practice 263
   for materials 264
   international 262
   national 262
statistics, *see* failure
   see quality
   see service life
   for materials selection 218
steel, *see* iron
   manufacture 56, 57
   properties 60
   stainless 58, 68
   uses 59, 67
   varieties 58, 67
strain
   effects on materials 93, 99, 149
   for metal hardening 40

measurement of 24
relation to stress 41
stress
   as cause of corrosion 144, 214
   as cause of cracking 82, 109, 149
   in materials 145, 149
   relaxation 152, 158
   tests 99
substitute materials, *see* past experience
   affected by production routes 221, 228
   non-interchangeability 219
   reasons for choosing 220, 225
   successes and failures 224
   to simplify a product 223
surface finishing, *see* electroplating
             see painting
   anodic coatings 70, 195, 201
   as a system 182
   conversion coatings 67, 78, 189, 200
   electroless deposition 199
   mechanical processes 192, 200
   plastisol dips 189
   processes for application 190, 191
   selection of finishing system 188, 193
   suitable substrates 182, 191
   surface hardening treatments 67, 78
   with engineering properties 78, 193, 202,
      204
temperature, *see* cryogenics
   affecting operation of product 28, 42,
      212
   as an environment 2
   deciding choice of materials 42
   effects on materials 58, 144, 155
   thermal expansion of plastics 28, 100
tensile properties, *see* strain
   see stress
   exploitation of 24
   of materials 101, 102
   testing for 40, 41
testing of materials
   accelerated methods 146, 159, 208
   chemical resistance 146
   electrical properties 158
   non-destructive 217, 218
   physical properties 99, 108
   strength 143
   variability of replicates 24, 40, 217

toughness, *see* fracture
  fracture and notch effects 24, 48, 61
  test methods 24
toxic fumes, *see* smoke emission
trade names 33
transparency
  ceramics 87
  glass 82
  plastics 112
  to non-visible light 237

value engineering 31
vapour deposition
  of inorganic compounds 201, 202
  of metals 201, 227, 232
  of plastics 203
  processes 202
vitreous enamel
  application processes 204
  materials 204
  performance 205
  substrates 205

wealth, creation of 1
wear, *see* ceramics
  applications 68, 80
  environmental effects 164

materials to resist 58, 82, 105, 130, 145,
  193, 199
  mechanical 105, 139, 238
  resistant coatings 193, 199, 202, 205,
  207
  test methods 150
weather, *see* environmental vectors
  effects on materials 109, 146, 151, 154
  effects on products 154
  humidity 2, 146, 151, 154
  radiation 2, 48, 154
  temperature 2, 152, 154
Weibull distribution, *see* statistics
  of test results 149
welding
  joint design 73, 215
  materials for welding 73
  processes 56, 235
  substrates
    metal 48, 56, 62, 73, 147, 221, 235
    plastic 119, 236
    other 238

zinc
  alloys 59, 71
  castings 59, 71
  supplies 55
  surface finish 188, 198
  world sources 55